STUDY GUIDE

CORE TOPICS

THIRD EDITION

Discover Biology

Michael L. Cain
Bowdoin College

Hans Damman
Duncan, British Columbia

Robert A. Lue
Harvard University

Carol Kaesuk Yoon
Bellingham, Washington

Contributing Author
Richard E. Morel
Signal Mountain, Tennessee

by

David Demers
Sacred Heart University

Edward Dzialowski
University of North Texas

Stephen Lebsack
Lynn-Benton Community College

W. W. NORTON & COMPANY
NEW YORK • LONDON

Printed in the United States of America

Third Edition

ISBN-10: 0-393-92845-4
ISBN-13: 978-0-393-92845-7

W. W. Norton & Company, Inc., 500 Fifth Avenue, New York, NY 10110
www.wwnorton.com

W. W. Norton & Company Ltd., Castle House, 75/76 Wells Street, London W1T 3QT

1 2 3 4 5 6 7 8 9 0

Contents

CHAPTER 1

The Nature of Science and the Characteristics of Life

Getting Started

Below are a few questions to consider prior to reading Chapter 1. These questions will help guide your exploration and assist you in identifying the key concepts presented in this chapter.

1. What is a hypothesis, and why can it never be proven to be true?
2. What is life, and where did it come from?
3. What are some of the characteristics that all living things have in common?
4. Are viruses, such as the influenza virus that causes the common flu, alive or not?
5. What is a biological hierarchy, and how does knowing about it help us in our everyday lives?
6. What role does the sun play in sustaining life on our planet? What are the differences between producers, consumers, and decomposers in a "food web"?
7. What is a nanobe, and how does it compare to a virus?

A Guide to the Reading

The following concepts typically give students the most difficulty when exploring the content in Chapter 1 for the first time. For each concept, one or more references have been identified that may help you gain a better understanding of these potentially problematic areas.

Limits of Science

Biologists employ the scientific method to study the living world. This process involves the observation of a phenomenon followed by the formation of a hypothesis. A testable prediction based on the hypothesis is then made, which is followed by experimentation to determine the validity of the hypothesis. While the scientific method provides scientists with a means to investigate the natural world, it cannot be used to investigate matters of faith or morality. Consider your own religious beliefs about the origin of life. Why is it that science cannot be used to investigate this question?

For more information on this concept, be sure to focus on:

- In Section 1.1, The Scientific Method Has Limits.

What Is Life?

Characteristics that are common to all living organisms help us to better define what makes something "alive." Seven properties are described in Chapter 1 that are common to all living things. (Recall that all living organisms [1] are built of cells, [2] reproduce themselves using the hereditary material DNA, [3] develop, [4] capture energy from their environment, [5] sense their environment and respond to it, [6] show a high level of organization, [7] evolve.) Consider examples of how you possess each of these characteristics.

For more information on this concept, be sure to focus on:

- Section 1.2, The Characteristics That All Living Organisms Share

Evolution vs. Adaptation

A group of organisms capable of breeding with one another is referred to as a *species.* Each new generation of a species has the potential to introduce change to certain characteristics possessed by members of the group. If such a change imparts a survival or reproductive advantage to individuals possessing the new characteristic, it can be called an *adaptation.* Evolution occurs when adaptations accumulate over time and change the characteristic(s) that define the species.

For more information on this concept, be sure to focus on:

- In Section 1.2, Living organisms evolve

The Hierarchy of Life

All organisms can be described using a biological hierarchy that spans a range, from the molecules that make up the most basic unit of life, the cell, to entire organisms and the environments in which they live. Many multicellular organisms, including humans, have arrangements of organs (composed of a number of different tissue types) that work together as a system to help sustain the individual. Individuals may then belong to a population, a large group of organisms of the same species. Populations may then live and interact with other populations in a living system referred to as a *community.* A collection of communities may then make up an ecosystem. The Earth (biosphere) can then be considered as a large, enclosed system containing many different biomes that consist of the various ecosystems on our planet.

For more information on this concept, be sure to focus on:

- Section 1.4, The Biological Hierarchy
- Figure 1.11, The Biological Hierarchy

Tying It All Together

Not surprisingly, concepts presented in this introductory chapter are revisited and discussed in greater detail in subsequent chapters, including:

Cell Theory of Life

- Chapter 5 – Section 5.1, Cells: The Simplest Units of Life

Energy Flow in a Living System

- Chapter 7 – Section 7.1, The Role of Energy in Living Systems
- Chapter 7 – Section 7.2, Using Energy from the Controlled Combustion of Food

DNA as Hereditary Material

- Chapter 12 – Section 12.1, The Search for the Genetic Material

Evolution

- Chapter 16 – Section 16.1, Biological Evolution: The Sum of Genetic Changes
- Chapter 16 – Section 16.5, The Impact of Evolutionary Thought
- Chapter 19 – Section 19.2, The History of Life on Earth

The Biological Hierarchy and Food Webs

- Chapter 20 – in Section 20.4, The biosphere can be divided into biomes
- Chapter 23 – in Section 23.1, Food webs consist of multiple food chains

Practice Questions

Factual Knowledge

1. The discovery of nanobes described at the beginning and end of the chapter highlights which of the following questions?
   a. Where does the energy that powers life come from?
   b. What is life and how many characteristics must an organism have to be considered alive?
   c. Has all life evolved from a single common ancestor?
   d. When does life begin and end?
   e. all of the above
2. A newly discovered microscopic structure is hypothesized to be a living organism. Which of the following characteristics would support the contention that this "organism" may be alive?
   a. It contains DNA.
   b. It is made of a single cell.
   c. It utilizes energy.
   d. It is highly organized.
   e. all of the above
3. Consider the structure labeled "lungs" in Figure 1.9 of your textbook. In terms of the biological hierarchy described in the chapter, which of the following structures are contained within this body part?
   a. molecules
   b. cells
   c. organ systems
   d. tissues
   e. More than one of the above is correct.
4. Advancing biological knowledge depends on rigorously testing hypotheses about the living world. (True or False)
5. Scientists begin constructing a simplified view of nature by doing experiments and forming theories. (True or False)
6. Which of the following would you expect to find if cells from a rose, a mushroom, and a cat were analyzed?
   a. DNA
   b. tissues
   c. viruses
   d. organs
   e. biospheres
7. Evolution represents biological change that accumulates through time. However, this change is not always apparent at every level in the biological hierarchy. Scientists most commonly observe the effects of evolution at the level of the ________.
   a. species
   b. community
   c. cell
   d. organism
   e. ecosystem
8. Like living organisms, nanobes are capable of reproducing on their own. (True or False)

9. If a cell contains DNA and other biological molecules, is highly organized, and shows signs of having once grown and developed, then it must currently be alive. (True or False)

10. Consider the studies conducted by Dr. Burkholder to determine the cause of the mysterious die-offs plaguing fish in the rivers of North Carolina described in the chapter. Her postulation that the deaths were caused by a type of organism called a *protist* is considered a(n) _______.
    a. experiment
    b. observation
    c. hypothesis
    d. theory
    e. test

11. Match each term with the best description.
    __ adaptation
    __ ecosystem
    __ development
    __ evolution
    __ multicellular
    __ hypothesis
    __ species
    a. organisms containing many different types of specialized cells
    b. a collection of organisms and the places where they live
    c. biological change accumulating in a species over time
    d. group of interbreeding individuals with fertile offspring
    e. features that provide a survival or reproductive benefit
    f. changes that occur within the life of an individual
    g. tentative, testable explanation for an observation

12. Which of the following is *not* a potential source of energy for a food web?
    a. plants
    b. animals
    c. the sun
    d. nonliving materials, such as iron
    e. All of the above are potential sources of energy.

Conceptual Understanding

1. Which of the following is *not* a scientifically testable hypothesis?
    a. The growth of peanut plants is affected by soil nitrogen concentration.
    b. El Niño–induced weather patterns influence fish harvests in the Pacific.
    c. Biological diversity is the result of a single divine creation.
    d. High dietary fiber leads to lower cholesterol in humans.
    e. Natural fires lead to the rejuvenation of certain plant communities.

2. Your television won't turn on, and you speculate that a circuit breaker in the house has been tripped. In scientific terminology, such reasoning would best be described as
    a. forming conclusions from the results of experiments.
    b. controlling variables in a repeated manipulation of nature.
    c. developing an observation based on a hypothesis.
    d. developing a hypothesis based on an observation.
    e. testing a prediction generated from a hypothesis.

3. Consider Dr. Burkholder's experiments on the mysterious die-offs of fish in North Carolina rivers. Which of the following best represents her main hypothesis?
    a. Fish maintained in laboratory environments were dying when exposed to river water.
    b. Microscopic organisms called *protists* accumulated in large numbers in laboratory fish tanks just before the fish died.
    c. Microscopic organisms called *protists* were responsible for causing the fish die-offs.
    d. Fish maintained in laboratory environments and not exposed to river water did not die.
    e. none of the above

4. Different organisms undergo different developmental processes. Consider the development that both humans and the monarch butterfly undergo, and match the corresponding stages listed below.
    __ fertilized egg
    __ fetus
    __ baby
    __ adult
    a. chrysalis
    b. butterfly
    c. egg
    d. caterpillar

5. Characteristics that are beneficial to species, such as the average speed at which pronghorn antelope can run, are known as evolutionary _______ and are determined largely by the _______ of an individual.
    a. changes; DNA
    b. adaptations; DNA
    c. dead ends; DNA
    d. hypotheses; energy
    e. food webs; energy

6. In most living systems, the energy that powers life comes ultimately from
   a. plants.
   b. animals.
   c. fungi.
   d. bacteria.
   e. the sun.

7. An important feature of most food webs is that energy
   a. ultimately comes from sources other than the sun.
   b. flows in one direction, from animals to plants and then to the sun.
   c. flows in one direction, from the sun to producers and then to consumers.
   d. flows in multiple directions, from plants to animals and then back again.
   e. seems to follow no particular pattern.

8. In a particular food web, the leaves of oak trees are eaten by moth caterpillars. Some of the caterpillars are then eaten by birds, which in turn are eaten by snakes. In this food web, the consumers are the
   a. oak trees.
   b. oak trees and caterpillars.
   c. caterpillars and birds.
   d. birds and snakes.
   e. caterpillars, birds, and snakes.

9. Which of the following was *not* shown by Stanley Miller's classic experiments on the origin of life?
   a. Biological molecules can form spontaneously in the proper chemical environment.
   b. Electricity is an energy source that can drive primordial chemical reactions.
   c. Methane, ammonia, and hydrogen can be used as primary ingredients in the primordial soup.
   d. Certain amino acids form spontaneously during the primordial soup reactions.
   e. Given time, the primordial soup eventually will yield living cells.

10. Consider the following list of characteristics common to all living organisms. Match the characteristics with the appropriate example.
    __ reproduction using DNA
    __ capturing energy from the environment
    __ sensing and responding to the environment
    __ displaying a high level of organization
    __ evolving
    a. Disruptions in the arrangement of internal organs can have fatal consequences for an organism.
    b. Pronghorns with the ability to run faster have a survival advantage over slower pronghorns.
    c. Some plants have the ability to undergo photosynthesis as well as capture and devour insects.
    d. Characteristics are passed from parents to offspring.
    e. Sunflowers typically turn toward the warmth of the sun.

11. Which of the following statements about nanobes is *false*?
    a. Most biologists have concluded that nanobes are not living.
    b. Nanobes exhibit no characteristics of living organisms.
    c. Nanobes possess cell-like structures, including membranes.
    d. Nanobes contain DNA.
    e. Scientists have observed the growth of nanobe colonies in the lab.

12. Viruses are essentially a particle of genetic information wrapped in a protein coat. Despite the fact that these particles are nonliving, they display certain characteristics of living organisms, including the ability to evolve. Which of the following is an example of the evolutionary capability of viruses?
    a. After receiving a vaccination against the flu one year, a person comes down with the flu the following year.
    b. Certain viruses may make use of RNA as genetic material.
    c. Viruses are incapable of reproducing on their own.
    d. Viruses are incapable of capturing energy from their environment.
    e. all of the above

Related Activities

- Consider the article "Baby's Fate in Summit Court" in the Biology in the News section at the end of this chapter. One day you may find yourself embroiled in a debate on what characteristics an individual must possess to be considered "alive." Do you consider the ability to respond to one's environment as a critical component of life? Is the ability to sustain oneself without artificial support important? Since science cannot be used to address such moral issues, it must be up to society to determine the definition of life. Explore the argument using the World Wide Web, and compose a one-page summary of the main arguments from both sides of the issue. After considering both sides, summarize your views on this controversial topic.
- Review the Biology Matters box, "How Much to Eat of What?" in Chapter 1. This passage describes the alarm-

ing trend of obesity in the United States and worldwide. In order to combat bulging waistlines, many Americans have resorted to specialized diets such as the Atkins diet or the South Beach diet. Using the World Wide Web, research a typical meal plan proposed by each of these diets and compare their nutritional makeup. How do these diets compare to the diet guidelines suggested in your textbook? Write a one-page summary of your findings.

- We live in a society in which it is no longer reasonable to believe that people can remain scientifically illiterate and still function well in the modern world. Review Section 1.4, The Biological Hierarchy. At the end of this section are several examples of how understanding this hierarchy is important for everyday life. Review each of the examples and conduct the following exercise: Using your local newspaper, a magazine, or the World Wide Web, identify an article that demonstrates humankind's interaction with the natural world at each of the levels of the hierarchy. Compose a one-page summary of your examples to show how understanding our relationship with the natural world can be beneficial to our everyday lives.

Answers and Explanations

Factual Knowledge

1. **b** The discovery of nanobes, just like the discovery of viruses, raises the question of what is inanimate and what is alive. Recall that there are several characteristics that are common to all living organisms. While nanobes possess many of these characteristics, they do not possess all of them, precluding their classification as a living organism.
2. **e** All of these characteristics are shared by all living organisms. However, to be considered "living," an organism must possess all of the characteristics described in the chapter.
3. **e** Recall that lungs are considered organs. In terms of the biological hierarchy, organs are built from a collection of tissues that contain cells assembled from molecules. Lungs would consequently be considered part of the respiratory organ system that enables us to breathe.
4. **True** Rigorous hypothesis testing is the core of scientific inquiry.
5. **False** The first steps in the scientific method are observation and hypothesis formation. Experiments are subsequently performed to test the hypothesis. The experimental results are then analyzed and used to support or reject the hypothesis. The term *theory* is reserved for well-tested and well-supported hypotheses.
6. **a** DNA is universal in all cells. While viruses may often infect cells, they are not typically present in healthy cells. The other choices are all structures larger than a cell.
7. **a** Evolution is first evident at the species level, mostly for reasons having to do with the inheritance of adaptations within a population.
8. **False** Recall that while nanobes have been demonstrated to contain DNA, they have not yet been observed to be capable of reproducing on their own.
9. **False** Consider that a dead cell would show all of the characteristics listed. In order for a cell to be considered alive, it must possess all of the characteristics described in the chapter, including energy utilization and the ability to respond to its environment.
10. **c** Recall that a hypothesis is a testable explanation for a given observation. In this case a test, or experiment, was used to determine the validity of the proposed hypothesis.
11. **e** adaptation
    b ecosystem
    f development
    c evolution
    a multicellular
    g hypothesis
    d species
12. **e** Recall that most food webs begin with the capture of sunlight energy by organisms that fall into the producer category, including most plants. Consumers, including most animals, may then prey upon both producers and other consumers for their own energy needs. Some biological systems devoid of light (such as the bottom of the ocean), may rely on energy derived from nonliving material for sustenance.

Conceptual Understanding

1. **c** Using divine creation as an explanation for biodiversity requires a faith-based hypothesis that an all-powerful deity exists. Science cannot work with such hypotheses because they are inherently untestable.
2. **d** This is the process of going from observation (the television won't turn on) to tentative explanation (hypothesis about the circuit breaker).
3. **c** This is the only choice that provides a possible and testable explanation for the fish die-offs. The other choices are simply observations about the phenomenon.

4. Refer to Figure 1.6 to compare the developmental process that both humans and the monarch butterfly undergo.
   c fertilized egg
   d fetus
   a baby
   b adult

5. **b** Evolution is caused by the accumulation of change within a species through time, and adaptations are the changes that provide survival or reproductive success. The majority of adaptations are products of an organism's DNA, which is inherited from the individual's parents.

6. **e** Most living systems ultimately derive their energy from the sun. However, there are examples, such as ecosystems present at the bottom of the ocean, that do not rely ultimately on the sun for energy.

7. **c** Refer to Figure 1.13, and note how the energy flows from the sun to producers and then to consumers of one kind or another.

8. **e** The oak trees (plants) are the producers; the remaining organisms in the scenario would be considered a type of consumer.

9. **e** Review Box 1.1. Miller's experiment showed that certain chemical precursors of life could form in the primordial soup under specialized conditions thought to resemble the environment on early Earth. His experiment did not yield any evidence that actual living cells would eventually develop from the chemical precursors.

10. **d** reproduction using DNA
    c capturing energy from the environment
    e sensing and responding to the environment
    a displaying a high level of organization
    b evolving

11. **b** Nanobes are similar to living organisms in that they possess many cell-like structures, possess DNA, appear to grow in the lab, and appear to be highly organized. However, it is not certain that nanobes have the ability to reproduce on their own or evolve.

12. **a** Since the influenza virus (the causative agent of the flu) changes or evolves each year, our body's immune system is incapable of recognizing the new forms that appear each year. This is why re-vaccination is required on an annual basis.

CHAPTER 2

Organizing the Diversity of Life

Getting Started

Below are a few questions to consider prior to reading Chapter 2. These questions will help guide your exploration and assist you in identifying the key concepts presented in this chapter.

1. It is believed that all life on earth was derived from a single common ancestor. How do scientists determine how groups of organisms are related to one another?
2. How has the use of DNA technology improved scientists' ability to determine evolutionary relationships?
3. What are shared derived features, and how do they differ from convergent features when determining evolutionary relationships?
4. What can the construction of an evolutionary tree tell us about the paternal instincts of dinosaurs?
5. How is the "wise man" (*Homo sapiens*) related to the "upright man" (*Homo erectus*) and the "handy man" (*Homo habilus*)?
6. Why is the mushroom on your slice of pizza more closely related to you than it is to the green pepper that sits next to it?

A Guide to the Reading

The following concepts typically give students the most difficulty when exploring the content in Chapter 2 for the first time. For each concept, one or more references have been identified that may help you gain a better understanding of these potentially problematic areas.

Common Ancestor

Scientists known as *systematists* work to study the relationships among different groups of organisms. A graphical summary of their work can take the form of an evolutionary tree. Just like a family tree, an evolutionary tree depicts relationships—in this case, relationships between different species or groups of organisms. Descendants from a single ancestral species are depicted on the evolutionary tree as the tip of a branch. The most recent common ancestor of two divergent species is therefore represented on the tree by the point where one lineage branch joins another. While two groups of organisms may have many common ancestors, there can only be one most recent common ancestor, represented by the fork on the tree.

For more information on this concept, be sure to focus on:

- In Section 2.1, Groups are related through their most recent common ancestor

Shared Derived Features vs. Convergent Features

Descendants from a common ancestor often share similar features. The unique features of an ancestral species (species that appeared first) that have been passed on to the divergent descendant species (species that appeared after the ancestral species) are referred to as *shared derived features*. Groups possessing such shared derived features are marked as a set of close relatives on an evolutionary tree. However, not all features shared among groups of organisms represent evolutionary relationships. Some features may have indeed evolved independently within two divergent groups of organisms. Such features are referred to as *convergent features*.

For more information on this concept, be sure to focus on:

- In Section 2.1, Examining shared derived features helps determine the most recent common ancestor
- In Section 2.1, Convergent features are similarities that do not indicate relatedness
- Figure 2.2, Shared Derived Features Define Evolutionary Relationships
- Figure 2.3, Misleading Convergent Features

Domains

In the chapter, the *Linnaean Hierarchy* is described as a system adopted by scientists to classify organisms into specific taxonomic groups reflecting their relationships with other organisms within the hierarchy. The original hierarchy described just two kingdoms: plants and animals. Over time, scientists have expanded this system to include upwards of eight kingdoms of organisms (your textbook adopts the widely used six-kingdom system). Beyond this, scientists have recognized a need to create even larger groupings of organisms, referred to as a *domain*. There are three proposed domains: *Bacteria*, *Archaea*, and *Eukarya*. When reviewing the chapter, be sure to pay particular attention to the key differences that distinguish the Bacteria from the Archaea and to how scientists have been able to use DNA to help support this proposed domain structure.

For more information on this concept, be sure to focus on:

- Section 2.3, A Classification System for Organizing Life: The Linnaean Hierarchy and Beyond

Horizontal Gene Transfer

In Chapter 2, the power of DNA technology to assist systematists in defining and refining evolutionary relationships between groups of organisms is highlighted with numerous examples. However, the study of DNA has also raised new questions. The text describes an example where studies revealed the presence of bacterial DNA within the cells of organisms from both the Archaea and the Eukarya domains. Dr. W. Ford Doolittle's theory of horizontal gene transfer was developed to explain how organisms that diverged so long ago could share the same bacterial DNA.

For more information on this concept, be sure to focus on:

- In Section 2.4, A surprising tangle can be found at the roots of the tree of life

Tying It All Together

Several concepts presented in this chapter are revisited and discussed in greater detail in subsequent chapters, including:

Kingdoms of Life

- Chapter 3 – Section 3.1, The Major Groups in Context

Modern DNA Technologies

- Chapter 12 – Section 12.1, The Search for the Genetic Material
- Chapter 15 – Section 15.3, Applications of DNA Technology

Evolution and the Origin of Species

- Chapter 16 – Section 16.1, Biological Evolution: The Sum of Genetic Changes
- Chapter 16 – Section 16.4, Strong Evidence Shows that Evolution Happens
- Chapter 19 – Section 19.2, The History of Life on Earth

Humankind's Closest Relatives

- Interlude D, Humans and Evolution

Practice Questions

Factual Knowledge

1. Which of the following could potentially be used by a systematist to help build an evolutionary tree?
   a. DNA
   b. structural similarities
   c. behavior
   d. shared derived characteristics
   e. all of the above

2. When the incredibly well-preserved specimen named Otzi the Iceman was first discovered in 1991, scientists feared that he might be an elaborate hoax. Which of the following lines of evidence finally convinced scientists that Otzi was real?
   a. birch fungus that was found with his body
   b. the clothing that was found on his body
   c. the numbers of tattoos on his body
   d. the weapons that were found on his body
   e. his DNA

3. Biologists called ________ study relationships among different groups of organisms. They use characteristics of the organisms they study to construct ________, which depict the divergence of species that have occurred over time.
   a. systematists; evolutionary trees
   b. genealogists; evolutionary trees
   c. systematists; Linnaean hierarchies
   d. genealogists; Linnaean hierarchies
   e. Linnaeans; shared derived features

4. Organisms that share physical traits are almost always members of the same species. (True or False)

5. Refer to Figure 2.2 in your textbook. Of the characteristics listed, which do salamanders and humans share?
   a. lungs and backbone
   b. claws or nails
   c. fur and mammary glands
   d. opposable thumbs
   e. all of the above

6. Again refer to Figure 2.2 in your textbook. Of the characteristics listed, which do chimpanzees and humans share?
   a. lungs and backbone
   b. claws or nails
   c. fur and mammary glands
   d. opposable thumbs
   e. all of the above
7. When groups of organisms share characteristics because they evolved the same features independently, rather than through common descent, we call these characteristics:
   a. systematic
   b. convergent
   c. ancestral
   d. derived
   e. evolutionary
8. Refer to the Tree of All Life depicted in Figure 2.9 in your textbook. Which of the following statements about this tree is *true*?
   a. Bacteria were the first organisms to evolve from the universal ancestor.
   b. Archaea evolved from the common ancestor of all eukaryotes.
   c. It is uncertain whether ciliates or diatoms evolved first.
   d. Fungi are more closely related to animals than they are to plants.
   e. all of the above
9. Which of the following is a prime example of why DNA technology has been used extensively by systematists to analyze evolutionary relationships?
   a. All organisms use DNA as hereditary material.
   b. DNA allows for direct comparisons between physically dissimilar groups of organisms such as bacteria, plants, and animals.
   c. DNA can be obtained from the fossil remains of extinct organisms.
   d. DNA can be analyzed from relatively "featureless" organisms.
   e. all of the above
10. DNA technology was the primary tool used to determine that the Archaea belonged to a different evolutionary group than the Bacteria. (True or False)
11. Match each term with the best description.
    __ horizontal gene transfer
    __ phylum
    __ evolutionary tree
    __ systematics
    __ shared feature
   a. pictorial representation of relatedness
   b. general study of evolutionary relationships
   c. a large group of closely related classes of organisms
   d. common characteristic of two or more groups
   e. unexpected movement of DNA between separated lineages
12. In the Linnaean hierarchy, what are the units of classification from smallest to largest?
   a. species, genus, order, family, phylum, class, kingdom
   b. kingdom, species, order, genus, family, class, phylum
   c. species, genus, family, order, class, phylum, kingdom
   d. phylum, family, genus, species, order, kingdom, class
   e. family, species, genus, order, kingdom, class, phylum
13. The name *Homo sapiens* reflects the species name and genus name for humans, in that order. (True or False)

Conceptual Understanding

1. Refer to the swallowtail butterfly evolutionary tree shown in Figure 2.1*b* of your textbook. Which of the following statements about this tree is *false*?
   a. Descendants are represented by the tips of the branches.
   b. The universal ancestor for all species is in the upper right of the diagram.
   c. Species A and B are the most closely related.
   d. Species D is more closely related to E than it is to A or B.
   e. The ancestor of A + B + C had features lacking in species E.
2. Refer to Figure 2.4 of your textbook, and then determine which of the following *living* organisms has the most shared, novel features in common with birds.
   a. lizards
   b. turtles
   c. crocodilians
   d. saurischian dinosaurs
   e. ornithischian dinosaurs
3. The opposable thumb shared between humans, chimpanzees, and pandas is a product of convergent evolution among these species. (True or False)
4. Recall the study from your textbook regarding HIV infection in patients of a single dentist. In this case, which of the following important lines of

systematic evidence were used as the key measure of relatedness?
   a. shared derived features
   b. convergent features
   c. similarities in DNA
   d. evolutionary trees
   e. all of the above

5. Which of the following would be considered a key medical consequence resulting from the discovery that fungi and animals are more closely related than previously thought?
   a. Antibiotics used to treat bacterial infections are ineffective against fungal infections.
   b. Yeast (a type of fungus) could hold the key to curing Lou Gehrig's disease.
   c. Fungal infections have traditionally been very difficult to cure since drugs that harm fungi typically have adverse effects in the body.
   d. The research of fungi may someday allow scientists to discover new treatments for several genetic diseases.
   e. all of the above

6. Which of the following observations would be explained by the hypothesis of horizontal gene transfer?
   a. Bacterial DNA has been found in the cells from organisms in the Eukarya domain.
   b. Genes may be passed horizontally between siblings spontaneously.
   c. Shared derived features may evolve independently in separate lineages.
   d. Organisms may be classified in more than one grouping in the Linnaean hierarchy.
   e. all of the above

7. In order for the dinosaurs to be considered a real group, many systematists say that the ________ must also be included.
   a. primates
   b. birds
   c. mammals
   d. fungi
   e. reptiles

8. The discovery of a fossil dinosaur named *Oviraptor* at its nest has confirmed the view of many paleontologists that some dinosaurs exhibited parental care. This is an example of
   a. the power of convergent evolution.
   b. the importance of using shared, derived features to make evolutionary trees.
   c. using evolutionary trees to predict the biology of organisms.
   d. how DNA information is critical to generating evolutionary trees.
   e. poor science.

9. Consider Figure 2.11 in your textbook. Based on this family tree, which primate's DNA would be most unlike that of a human?
   a. lemur
   b. spider monkey
   c. gibbon
   d. orangutan
   e. gorilla

10. You have been named head of a scientific team investigating the cause of the infamous Irish potato famine discussed in the Biology in the News article at the end of the chapter. To determine if the *Phytophthora infestans* strain known to be responsible for the famine is the same as the US-1 strain described in the article, you would
   a. compare the physical dimensions of the two strains.
   b. test the US-1 strain on Irish potato crops.
   c. observe the growth of *Phytophthora infestans* strains in the laboratory.
   d. compare the DNA from the two strains.
   e. determine the most common ancestor.

11. Fungi are more closely related to plants than they are to animals. (True or False)

12. Flight is a characteristic shared between organisms from different groups. Both birds (Class Aves) and bats (Class Mammalia) are vertebrates that have the ability to use their front limbs to fly. Thus, flight is an example of
   a. a shared, derived feature.
   b. a Linnaean feature.
   c. convergent evolution.
   d. horizontal gene transfer.
   e. There is not enough information to answer the question.

13. If two organisms belong to the same class in the Linnaean classification hierarchy, it is also true that these organisms belong in
   a. the same genus and species.
   b. the same kingdom and phylum.
   c. the same family.
   d. the same order.
   e. a different kingdom and phylum.

Related Activities

- At first glance, the birds and the dinosaurs may not appear very similar. After all, how could the fearsome *Tyrannosaurus rex* be related to a pigeon? Upon closer examination, however, the common features between the two groups are striking. Find a resource other than your textbook that compares dinosaurs to birds. What do the groups have in common? What are the differences? What do scientists still need to know to understand more clearly the evolutionary relationship between birds and dinosaurs?

- Look ahead in your textbook to Interlude D on human evolution. Skim through the section and list as many primate or human characteristics you can find that would be useful for assembling a human evolutionary tree. Explain why the traits you have listed may have value in reconstructing human evolution.
- Using the World Wide Web, retrieve some basic information about the legendary creature Sasquatch (Bigfoot). Compare the reported biological characteristics of Sasquatch to those of modern humans. Now assume that Sasquatch actually becomes available for scientific study. Make a list of the information you would want to gather from the specimen for use in determining its evolutionary affinity to humans. You may wish to refer to the study of Otzi the Iceman discussed at the beginning and end of the chapter.

Answers and Explanations

Factual Knowledge

1. **e** All of these could be used to help build an evolutionary tree.
2. **e** DNA comparisons showed that Otzi the Iceman was closely related to Europeans and therefore could not be a hoax involving a mummified human taken from some other location. All the other clues, while useful anthropologically, did not identify conclusively the ancestry of the Iceman.
3. **a** Recall that when constructing evolutionary trees, systematists work to understand how groups of organisms are related.
4. **False** Recall from the text that DNA technology provides a powerful tool for scientists to determine evolutionary relationships. An unknown organism that physically resembles a known species may in fact contain distinct DNA, indicating that it is likely a different species.
5. **a** Salamanders and humans share only those features that arose before the salamanders branched off from the ancestral vertebrate lineage (i.e., a backbone and lungs). All the other features listed are derived characteristics that appeared in vertebrates after the salamander lineage diverged.
6. **e** The common link between chimps and humans evolved after all the traits listed had arisen in the vertebrate lineage. Thus the ancestral link possessed all of these traits and passed them on to both chimps and humans.
7. **b** Convergent evolution is a process that can produce similar characteristics in organisms as a solution to environmental challenges, rather than through common descent. It can thus be a problem for systematists trying to determine evolutionary relationships.
8. **e** All of these are correct interpretations of the information in Figure 2.9.
9. **e** All of these are examples of how DNA technology has been used by systematists to assist in determining evolutionary relationships.
10. **False** Recall from the text that DNA studies initially alerted scientists to the existence of the Archaea domain.
11. **e** horizontal gene transfer
    c phylum
    a evolutionary tree
    b systematics
    d shared feature
12. **c** Recall the catch phrase *K*ing *P*hillip *C*leaned *O*ur *F*ilthy *G*ym *S*horts to help you remember the correct order.
13. **False** In a scientific name the first name is the genus (*Homo*) and the second is the species (*sapien*).

Conceptual Understanding

1. **b** The universal ancestor would have existed in the distant evolutionary past, which is represented by the lower left corner of the diagram.
2. **c** Although birds are depicted to share a recent common ancestor with the Saurischian and Ornithischian dinosaurs, since these organisms are extinct, the crocodile would be the most closely related organism to birds in the figure.
3. **True** This is a classic example of convergent evolution in which the environmental benefits of an opposable thumb have forced common solutions to similar problems faced by the ancestors of these species.
4. **c** As is true in many cases, DNA evidence is a powerful tool for determining evolutionary relatedness. Refer to Box 2.1, Science Toolkit, to review the details of this interesting case.
5. **e** The realization that fungi are more closely related to animals than to plants has allowed scientists to discover new medically important links between these two groups. Refer to Figure 2.9 to see how these groups are located on the tree of life and Figure 2.10 for an example of a genetic disease shared between humans and fungi.

6. **a** The discovery that DNA from certain organisms can appear as a grab bag of DNA from across the tree of life has been explained by the horizontal gene transfer theory.

7. **b** There is now compelling evidence to suggest that the birds are direct descendants of the dinosaurs. Refer to Figure 2.4 for additional information on this discovery.

8. **c** Because birds and certain reptiles show parental care and because both are related to dinosaurs by virtue of other shared, novel features, a prediction can be made that dinosaurs likely also showed parental care.

9. **a** Humans and lemurs are distantly related within the primate family tree, since their common ancestor goes the furthest back in time (i.e., lower left of diagram). Thus, of the primates pictured, lemur DNA would be most unlike human DNA.

10. **d** While some of the approaches listed might be useful, the quickest and most reliable way to assess relatedness is through the use of DNA comparisons with known species.

11. **False** Using DNA techniques, fungi have recently been found to be more closely related to animals than to plants.

12. **c** Flight is an example of convergent evolution in birds and bats as this feature would have evolved independently in both groups.

13. **b** If two organisms are in the same class, then the only thing that can be said for certain is that they are in the same kingdom and phylum. Refer to Figure 2.6 to see these relationships.

CHAPTER 3 Major Groups of Living Organisms

Getting Started

Below are a few questions to consider prior to reading Chapter 3. These questions will help guide your exploration and assist you in identifying some of the key concepts presented in this chapter.

1. What can the most spectacular orchid discovery of the past 100 years teach us about biodiversity?
2. Why are the Greek terms *eu* ("true"), *pro* ("before"), and *karyote* ("kernel") important for understanding the three domains of life?
3. If a single bacteria of *E. coli*, normally found in the gut of humans, was allowed to reproduce without restriction, how long would it take to produce a population of 16,000,000 bacteria?
4. What is red tide, how does it affect humans, and which group of organisms is responsible for causing it?
5. What was the evolutionary innovation that enabled plants to grow to the towering heights seen among the trees?
6. Truffles, which can sell for upward of $600 per pound, are a member of which Linnaean kingdom?
7. What percentage of the roughly 1,000,000 known animal species are members of the animal group collectively known as the *insects*?

A Guide to the Reading

The following concepts typically give students the most difficulty when exploring the content in Chapter 3 for the first time. For each concept, one or more references have been identified that may help you gain a better understanding of these potentially problematic areas.

Prokaryotes vs. Eukaryotes

As you learned in Chapter 2, organisms can be classified according to several schemes. One such scheme categorizes organisms based upon the structure of their cells. This scheme distinguishes cells based on the presence (*eukaryote*) or absence (*prokaryote*) of a compartmentalized genetic center, or nucleus. Keep in mind that despite the similarity between the term *eukaryote* and the domain name *Eukarya*, these refer to different levels of classification. The term *eukaryote* can be used to describe the structure of an organism's cells (nucleus is present), while the domain *Eukarya* refers to the group of organisms represented by four kingdoms (Protista, Plantae, Fungi, and Animalia).

For more information on this concept, be sure to focus on:

- In Section 3.1, Organisms can also be identified as prokaryotes or eukaryotes

Prokaryote Abundance

In terms of the total number of organisms occupying the biosphere, the prokaryotes definitely take the cake. These tiny, microscopic, ancient organisms are found in nearly all locations on our planet and make up the vast majority of life. However, most of us remain unaware of their very existence since they cannot be seen with the naked eye. Scientists estimate there are roughly 5×10^{30} prokaryotes on earth! Part of the reason for the success of the prokaryotes is their ability to reproduce quickly. The bacteria *E. coli* is capable of doubling its population in as little as 20 minutes! If left to grow overnight, a single bacterium of this species could potentially produce a population of over 16 million individuals! Another reason for the success of the prokaryotes lies in their ability to utilize a variety of nutritional sources, acting as consumers, decomposers, and producers within a single food web.

For more information on this concept, be sure to focus on:

- In Section 3.2, Prokaryotes represent simplicity translated into success
- In Section 3.2, Prokaryotes exhibit unmatched diversity in methods of obtaining nutrition

The Protists—Evolution of the Eukaryote

The kingdom Protista is considered the most ancient group amongst the Eukarya domain. As discussed in the chapter, this group is extremely diverse in terms of size, shape, and lifestyle. As such, it is a group that has given scientists great difficulty in determining evolutionary relationships among its members. This struggle is illustrated by the evolutionary tree presented in Figure 3.7. Among the major groups of protists, only the *Diplomonads* are generally agreed to have branched off first. The remaining groups are all depicted to have branched off at roughly the same time, reflecting the fact that the order in which these groups evolved is still poorly understood by scientists. One thing is for certain, however—the protists were one of the first groups to develop organelles, tiny specialized compartments. It is believed that organelles evolved through the process of larger prokaryotic cells engulfing smaller prokaryotic cells, with the smaller, engulfed cells taking on specialized functions within the larger host cell. While organelles lost the ability to survive as a free-living organism, the host cell would become dependant on the function carried out by the organelle, resulting in a mutual relationship between the previously separate organisms and the generation of a new species.

For more information on this concept, be sure to focus on:

- In Section 3.3, Protists represent early stages in the evolution of the eukaryotic cell
- Figure 3.7, The Protista

Angiosperms and Flowers

As discussed in the chapter, the development of flowers is a relatively recent evolutionary event within the plant kingdom. Flowering plants (angiospersms) produce seeds that are characteristically surrounded by a protective tissue which may take the form of a type of fruit. What defines an angiosperm, however, is the flower—a structure used for reproduction. What may be hard to understand is that a flower typically contains *both* male and female reproductive structures as illustrated in Figure 3.11. Male reproductive structures take the form of the stamen, which consists of two parts: the filament and the anther. The anther is capable of producing pollen (male gametes; analogous to sperm in humans), which can be carried by either wind or animal to a nearby flower of the same species. The female reproductive structure, typically in the center of the flower, is called the *carpel* and consists of the stigma (site of pollen attachment), the style (carries pollen), and the ovary (analogous to the female egg in humans). When pollen reaches an ovary, the result is a plant embryo that can take the form of a seed. Since both female and male reproductive structures are present in the same flower, this means that an angiosperm is capable of self-pollination, or mating with itself.

For more information on this concept, be sure to focus on:

- In Section 3.4, Angiosperms produced the world's first flowers

The Fungi, a Group with Many Roles

As discussed in the chapter, fungi are described as playing many key roles in most ecosystems. The ability to play different roles stems from this group's ability to obtain nutrients and energy from a variety of sources and in a variety of ways. Fungi are capable of serving as a decomposer in an ecosystem by feeding upon dead or dying organisms. Their role is to return vital nutrients back to the ecosystem for use by the system's producers. In a similar fashion, some groups of fungi may specialize in living on other organisms. This relationship can either be beneficial (mutualism) or detrimental (parasitism) to the host. Keep in mind that these relationships may or may not be based upon the supply of nutrients. Therefore, there is no correlation between the terms *decomposer*, *consumer*, and *producer* and the terms *mutualist* and *parasite*.

For more information on this concept, be sure to focus on:

- Section 3.5, The Fungi: A World of Decomposers

Animal Body Cavities

One of the evolutionary innovations that helped animals develop the extreme diversity displayed by this group was the formation of a complete body cavity. The body cavity is defined as an interior space with openings at either end (mouth, anus). The text describes two distinct evolutionary lineages with a complete body cavity: the protostomes and deuterostomes. These two groups are distinguished by which of the two openings develops into the adult's mouth. In the protostomes (insects, worms, snails), during development the opening that will become the mouth forms first. In the deuterostomes (echinoderms and vertebrates—including humans) the opening that becomes the adult anus forms first during the organism's development. This distinction is important to understand as we learn more about organismal growth and development in subsequent chapters.

For more information on this concept, be sure to focus on:

- In Section 3.6, Animals evolved complete body cavities

Tying It All Together

Several concepts presented in this chapter are revisited and discussed in greater detail in subsequent chapters, including:

Eukaryotic Cell Structure

- Chapter 5 – Section 5.4, The Specialized Internal Compartments of Eukaryotic Cells

Photosynthesis

- Chapter 8 – Section 8.2, Photosynthesis: Capturing Energy from Sunlight

Chloroplasts

- Chapter 8 – in Section 8.2, Chloroplasts are the sites of photosynthesis

Biodiversity

- Chapter 20, The Biosphere

Practice Questions

Factual Knowledge

1. Most biologists currently estimate the total number of species that exist on Earth to be somewhere between ________ million.
   a. 1 to 2
   b. 2 to 3
   c. 3 to 30
   d. 30 to 100
   e. 100 to 500
2. Consider the evolutionary tree of life shown in Figure 3.1 of your textbook. According to this depiction, which group has evolved most recently?
   a. archaea
   b. ciliates
   c. diatoms
   d. animals
   e. plants
3. Which of the following is *not* found in the kingdom Bacteria?
   a. organelles
   b. DNA
   c. organized cell structure
   d. ability to reproduce
   e. ability to use energy
4. There are a number of reasons for the evolutionary success of the kingdoms Bacteria and Archaea. Which of the following characteristics allow these prokaryotes to be so successful?
   a. rapid reproduction
   b. diverse methods of obtaining nutrients
   c. the ability to survive extreme conditions
   d. the ability of some to live with oxygen and others to live without oxygen
   e. all of the above
5. Bacteria play many important roles in the biosphere and human society. One of these roles is to
   a. decompose the bodies of dead organisms.
   b. assist with digestion in animal bodies.
   c. make nitrogen available to plants.
   d. help clean up oil spills.
   e. all of the above
6. Match the kingdom Protista-related term with the best description.
   __ eukaryotic cells
   __ plantlike protists
   __ animal-like protists
   __ fungal-like protists
   __ sexual reproduction
   a. algae
   b. first appeared in early protists
   c. slime molds
   d. evolved when bacteria engulfed other bacteria
   e. ciliates

7. The evolutionary record suggests that the first occurrence of multicellularity was in the kingdom
   a. Archaea.
   b. Protista.
   c. Fungi.
   d. Plantae.
   e. Animalia.

8. The common characteristic of red tides, malaria, and potato blight is that they are all
   a. caused by bacteria.
   b. caused by protists.
   c. caused by fungi.
   d. diseases of plants.
   e. none of the above

9. Extensive root systems and the presence of a waxy cuticle are among the reasons that plants have become so successful on land. (True or False)

10. Consider the photographs of various plants in Figure 3.8 of your textbook. Which of the following statements about these organisms is *false*?
    a. They are prokaryotic photosynthesizers.
    b. They are multicellular photosynthesizers.
    c. Several produce flowers and seeds.
    d. They function as producers in food webs.
    e. Each has a waxy cuticle.

11. Which one of the following is *not* found in both the gymnosperm and angiosperm plants?
    a. well-developed vascular systems for internal transport
    b. extensive root systems for acquiring water and nutrients
    c. seeds that disperse from the parent and thus increase offspring survival
    d. sexual reproduction using special structures called flowers
    e. products that are useful for human consumption

12. Fungi are important to food webs because they
    a. photosynthesize.
    b. have prokaryotic cell structure.
    c. serve as key decomposers of dead organic matter.
    d. generate antibiotic chemicals that kill competing plants and animals.
    e. form mutually beneficial relationships with other organisms.

13. Consider the fungi shown in Figures 3.13 through 3.17 of your textbook. What do all of these organisms have in common?
    a. a mycelium
    b. many individual hyphae
    c. partial separation of adjoining cells
    d. spore production
    e. all of the above

14. Of the animal groups listed below, which one evolved first?
    a. flatworms
    b. sponges
    c. arthropods
    d. vertebrates
    e. molluscs

15. Because of their ingestive mode of nutrition, animals function as ________ in ecosystems, but never as ________.
    a. consumers and decomposers; producers
    b. producers; consumers and decomposers
    c. producers and consumers; decomposers
    d. decomposers; producers and consumers
    e. decomposers and producers; consumers

16. Which of the following animals shows an embryonic developmental pattern that classifies it as a deuterostome?
    a. birds
    b. snails
    c. insects
    d. worms
    e. none of the above

Conceptual Understanding

1. A researcher has just discovered what she thinks is a new species. Initially, all she knows is that it is eukaryotic. To which of the kingdoms below could this new organism possibly belong?
   a. Animalia
   b. Protista
   c. Fungi
   d. Plantae
   e. any of the above

2. Further analysis reveals the organism in question 1 is also multicellular and relies on other organisms for a source of carbon. Which kingdom(s) can now be *excluded* for this organism?
   a. Animalia
   b. Protista
   c. Fungi
   d. Plantae
   e. more than one of the above

3. Refer to Figure 3.1 in your textbook. Which of the following traits do the groups labeled *Bacteria* and *Animalia* have in common?
   a. They belong to the same domain.
   b. They belong to the same kingdom.

c. They share a common ancestor in the very distant evolutionary past.
d. They share a common ancestor in the very recent evolutionary past.
e. none of the above

4. Of all organisms, bacteria thrive in the broadest range of habitats. (True or False)

5. Refer to Figure 3.4 in your textbook. The cyanobacteria shown in this photograph are similar to plants in that they
a. photosynthesize.
b. can produce seeds.
c. have flowers.
d. live in very salty environments.
e. can be used to brew beer.

6. Eukaryotic cell organelles first emerged
a. from a specialized lineage of cells within the kingdom Protista.
b. when prokaryotes engulfed each other and became interdependent.
c. when bacteria made their first attempts at reproduction.
d. when multicellularity arose.
e. just before the origin of the animal and fungal kingdoms.

7. The protists make up a relatively small kingdom of prokaryotic organisms whose evolutionary relationships to the fungi, plants, and animals are well understood. (True or False)

8. Both plantlike protists and plants
a. are prokaryotic.
b. have some cellular organelles in common.
c. serve as decomposers in an ecosystem.
d. evolved after the first animals appeared.
e. were among the last multicellular groups to evolve.

9. Vascular systems were a key innovation in plant evolution because they
a. allowed plants to become taller and more efficient at nutrient transport.
b. made possible the production of oxygen during photosynthesis.
c. increased the success rate of sexual reproduction.
d. were needed before photosynthesis could evolve.
e. all of the above

10. The success of the angiosperms is due largely to an evolutionary innovation called the *flower*, which dramatically increases the efficiency of sexual reproduction. (True or False)

11. Select all of the statements below that correctly apply to the kingdom Fungi.
a. The fungi are eukaryotic and multicellular and rely on other organisms for food.
b. Some fungi form beneficial interrelationships with plants.
c. The fungal life cycle typically includes a spore stage.
d. Certain fungi are natural sources of antibiotic substances.
e. Fungi can form close associations with protists.

12. Which of the following natural processes would probably be most severely impacted if all fungi suddenly disappeared?
a. decomposition
b. photosynthesis
c. bacterial reproduction
d. viral reproduction
e. evolutionary innovation

13. Evolutionarily speaking, to which of the following animal groups are you most closely related?
a. starfish
b. flatworms
c. snails
d. birds
e. insects

14. Jellyfish and people are both animals, and so they share a number of key biological features characteristic of the entire kingdom. However, because of their distant ancestry, these two groups also exhibit important differences in body structure and function. Which of the following characteristics is present in humans but *not* in jellyfish?
a. true tissues
b. multicellular organization
c. organ systems
d. ingestion of food
e. mobility

15. All but the simplest animal groups (sponges) have ________ organization in their bodies. However, among the more advanced animals there are two distinct ways that the basic body tissues become arranged during the development of a fertilized egg. In ________, such as insects and earthworms, the first opening to form in the embryo becomes the mouth, while in ________, such as humans and sea stars, the mouth forms from the second embryonic opening.
a. cell-level; protostomes; deuterostomes
b. cell-level; deuterostomes; protostomes
c. tissue-level; protostomes; deuterostomes

d. tissue-level; deuterostomes; protostomes
e. organ-level; protostomes; deuterostomes

16. Outside of the animal kingdom, our closest evolutionary relatives are probably in the kingdom Protista. (True or False)

Related Activities

- Choose two examples of organisms from different kingdoms. List as many key biological features that the two have in common as you can think of. (Hint: Think back to the ideas discussed in Chapter 1.) Then expand your list to include the characteristics that are different about these organisms.
- Imagine that an elementary school child comes to you with a question about biology. As part of an assignment about the different kinds of organisms, he must classify as many types of living creatures as he can. Your friend is having trouble figuring out what to do with the lichens and viruses. Compose a simple explanation of how these two types of organisms are viewed in the Linnaean classification hierarchy.
- Consider the description of the Mimivirus in the Biology in the News section at the end of the chapter. This article illustrates how new discoveries are made by scientists on a regular basis. Explain how you would proceed if asked to attempt to classify this new discovery in the hierarchy of life. What would you attempt to determine first about the Mimivirus? What additional analyses could be performed to help classify the Mimivirus?

Answers and Explanations

Factual Knowledge

1. **c** This wide range of estimation indicates that scientists are constantly discovering new species and are far from knowing exactly how many species exist on our planet.

2. **d** The animals are in the part of the diagram that is farthest away from the universal ancestor, indicating that their lineage separated relatively recently from the closest non-animal ancestor.

3. **a** Bacteria are prokaryotes, which by definition means that they lack internal organelles. All the other choices are universal properties of life.

4. **e** Bacteria and Archaea take advantage of all of these characteristics to allow them to be successful.

5. **e** The overall diversity of bacteria encompasses a wide range of functions and lifestyles. All of the processes listed involve bacterial activity.

6. **d** eukaryotic cells
a plantlike protists
e animal-like protists
c fungal-like protists
b sexual reproduction

7. **b** While some degree of multicellularity is found in all kingdoms except Bacteria and Archaea, the kingdom Protista is the most primitive and has the fewest multicellular species, suggesting that it is the ancestor of the other multicellular eukaryotes.

8. **b** All three of these conditions are caused by protists.

9. **True** The root system provides water and nutrients, and the waxy cuticle helps prevent desiccation.

10. **a** Although they are indeed photosynthesizers, plants are not prokaryotic.

11. **d** Flowers are found in the angiosperms only. Instead of flowers, gymnosperms form cones or conelike structures for protection of their seeds.

12. **c** The first two choices are not correct for fungi, which are eukaryotes that cannot carry out photosynthesis. The last two choices, while true of fungi, are not what make fungi particularly important in food webs. The main role of fungi is decomposition.

13. **e** All of these are general features of fungal biology.

14. **b** Sponges are the most primitive animal group, being a relatively loose association of specialized cell types, which lacks even tissue-level organization.

15. **a** Producers capture energy directly from inorganic sources (e.g., the sun) and convert it to organic energy, as the plants in a food web do during photosynthesis. The animals in a food web gather premade organic energy from other organisms, in the role of either consumers or decomposers.

16. **a** Recall that in deuterostomes, the second opening that forms at either end of the body cavity eventually develops into the mouth. This group includes all vertebrates, such as the bird.

Conceptual Understanding

1. **e** All the kingdoms listed contain eukaryotic organisms.

2. **d** Only the plants can be excluded, since they are photosynthetic and produce their own food. The

other three kingdoms all contain at least some multicellular (or colonial) species that must rely on other organisms for their carbon source.

3. **c** The common connection between the bacterial and animal lineages occurs very early in the evolutionary history of life (near the left side of the diagram).

4. **True** Recall that habitat preferences in bacteria are diverse, including extremes of heat and cold, acidity, and salt concentration.

5. **a** Cyanobacteria, like plants, are capable of photosynthesis. The other answers are true of only one of the two organisms in question.

6. **b** Eukaryotic cells are hypothesized to have originated when one ancient prokaryote engulfed another. The other answers are not correct because the events either took place before organelles emerged (bacterial reproduction) or involved organisms that already had eukaryotic cells.

7. **False** Recall that protists are eukaryotic and that the evolutionary relationship between the protists and the other eukaryotic kingdoms is poorly understood.

8. **b** Plants and plantlike protists are both eukaryotic, so similar cellular organelles are to be expected. The decomposers of an ecosystem are typically fungi. Recall that protists represent an early stage in eukaryotic and multicellular evolution, so the last two answers are also incorrect.

9. **a** Plant vascular tissue makes long-range transport of nutrients and water possible. The plant body can thus become larger. The other choices are incorrect because the more primitive nonvascular plants also photosynthesize and reproduce sexually with great success.

10. **True** The evolutionary innovation of using flowers for reproduction was so successful that it caused an explosion of new species during plant evolution.

11. **b, c, d, e** Recall that some fungi are unicellular (e.g., yeast), and all of them absorb food produced by other organisms. The close associations formed with plants (mycelium) and protists (lichens) prove beneficial for both groups.

12. **a** Decomposition of organic material and the return of key nutrients to the nonliving portion of the biosphere are the two main roles that fungi play.

13. **d** Humans are vertebrates, so we can eliminate the starfish, flatworms, snails, and insects from the list. The only other vertebrates on the list are the birds.

14. **c** Recall that the evolutionary development of organs and organ systems followed the development of tissues on the evolutionary tree. Cnidarians (the jellyfish) do not contain organs or organ systems. All the other choices are found in both jellyfish and humans.

15. **c** All animals have cell-level organization, but the sponges lack organ- or tissue-level structure. Every animal more advanced than sponges has tissue-level organization. Recall that the terms *protostome* and *deuterostome* refer to the order of formation of the two openings on either side of the body cavity in animals. In protostomes, the first opening develops into the mouth whereas the second opening forms the mouth in deuterostomes.

16. *False* Recall that recent DNA evidence suggests a closer evolutionary relationship exists between the fungal and animal kingdoms.

INTERLUDE A Biodiversity and People

Getting Started

Below are a few questions to consider prior to reading Interlude A. These questions will help guide your exploration and assist you in identifying some of the key concepts presented in this interlude.

1. Why do scientists believe that humans are responsible for the new mass extinction?
2. Which kingdom holds almost half of the 1.5 million known species?
3. What is the significance of the barking deer discovery?
4. What earned the Xerces Blue butterfly its dubious fame?
5. Which organism has people in Guam fearful of the toilets they use?
6. How many pounds of solid waste are produced by the average college student each year?
7. Why was Chicken Little (". . . the sky is falling!") right about his prophetic warning?
8. What is an ecotron?
9. Why does the destruction of redwood forests contribute to the current water shortage in northern California?
10. Where do the anti-cancer drugs Taxol and Vinblastine come from?

A Guide to the Reading

The following concepts typically give students the most difficulty when exploring the content in Interlude A for the first time. For each concept, one or more references have been identified that may help you gain a better understanding of these potentially problematic areas.

Number of Species on Earth

While scientists have been hard at work classifying and naming the 1.5 million *known* species on earth, the *total* number of species living on earth can only be estimated. Because there is no direct way to count all of them, most estimates for the total number of species range anywhere from 3 to 30 million. These estimates are obtained in a number of different ways, such as the "fogging" technique employed by Terry Erwin as described in the text. It is important to understand that regardless of the technique used, these estimates all depend on a set of assumptions, which, of course, could always turn out to be incorrect. Therefore, the total number of species present on earth will always remain a controversial topic in science.

For more information on this concept, be sure to focus on:

- In Interlude A, Scientists use indirect methods to estimate total species numbers

Mass Extinction

As discussed in the text, several mass extinction events have occurred throughout the history of the earth. Many scientists believe that we are currently undergoing another mass extinction event driven by the presence of humans and the explosion in the human population. While it is difficult to determine the exact extinction rate occurring around the globe, scientists have documented that upwards of 20 percent of the freshwater fish and bird species that existed 2000 years ago are now extinct. The primary difference between this mass extinction event and those that occurred previously is the influence of the human element.

For more information on this concept, be sure to focus on:

- In Interlude A, The current mass extinction probably began with early humans
- In Interlude A, Scientists have made estimates of species losses occurring today

Competition Among Species

One of the major threats to biodiversity discussed in the text involves the introduction of foreign species to an ecosystem. This phenomenon occurs almost exclusively through the action of humans, either purposely or by accident. When such a foreign species is particularly well suited to an environment, it may thrive and begin to compete with existing species within the locale. In this case, the foreign species may actually deprive native species of resources, resulting in the extinction of the native species. The text describes several examples along with the impact such a loss of biodiversity has on an ecosystem.

For more information on this concept, be sure to focus on:

- In Interlude A, Introduced foreign species can wipe out native species
- Figure A.8, The Threat of Nonnative Species

The Effects of Mutation

The text describes the alarming threat posed by changes in the climate to biodiversity. One group that appears to be particularly affected by this threat is the amphibians. While many factors appear to be involved in the disappearance of amphibian species worldwide, one factor warrants particular attention: ultraviolet radiation. Due to the reduction in the amount of rain in the mountainous northwest United States, developing frog and toad eggs receive greater exposure to ultraviolet light. The text indicates this amplified exposure results in increased vulnerability to disease and infection. Why is this? It turns out that ultraviolet light is a powerful mutagen—an agent capable of altering the genetic code of life, DNA. When such mutations occur, genes vital to the survival of the organism may be rendered useless, resulting in a weakened individual. The concept of mutation will be covered in more detail in future chapters.

For more information on this concept, be sure to focus on:

- In Interlude A, Some threats are difficult to identify or define

Why Is Biodiversity Important?

While the loss of a single spider or frog species may seem insignificant in the grand scheme of things, it is important to understand the importance of biodiversity to the health of the ecosystems on our planet. The text describes research conducted by scientists that supports the notion that the greater the diversity of species present in an ecosystem, the heartier it is. This can be calculated experimentally using measures such as "productivity" (biomass produced). In addition, the biodiversity present in an ecosystem can help reduce invasion by foreign species and the potentially devastating competition that may result (see "Competition Among Species," above). Furthermore, humans depend on a wide variety of goods and services provided by the vast number of species in the biosphere. We are still discovering new and potentially beneficial products obtained from plant and animal species alike. Wouldn't it be a shame if the extinction of a particular plant species occurred prior to the discovery of its beneficial use as a new anti-cancer drug?

For more information on this concept, be sure to focus on:

- In Interlude A, Biodiversity can improve the function of ecosystems
- In Interlude A, Biodiversity provides people with goods and services
- Figure A.13, Biodiversity and Your Health

Tying It All Together

Several concepts presented in this chapter build upon concepts presented in previous chapters and may also be revisited and discussed in greater detail in subsequent chapters, including:

Mutation

- Chapter 17 – Section 17.4, Mutation: The Random Production of New Alleles

Adaptation to Environment

- Chapter 18 – Section 18.2, Adaptation Does Not Craft Perfect Organisms
- Chapter 18 – Section 18.4, Speciation: Generating Biodiversity

Mass Extinction

- Chapter 19 – Section 19.4, Mass Extinctions: Worldwide Losses of Species

Biodiversity

- Chapter 19 – Section 19.7, An Overview of the Evolutionary History of Life

Practice Questions

Factual Knowledge

1. Which of the following shows the proper order of groups from that with the most known species to that with the fewest?
   a. insects; non-insect animals; prokaryotes; plants; fungi; protists
   b. plants; fungi; protists; insects; non-insect animals; prokaryotes
   c. insects; non-insect animals; plants; fungi; protists; prokaryotes
   d. plants; fungi; protists; prokaryotes; insects; non-insect animals
   e. fungi; plants; prokaryotes; protists; non-insect animals; insects

2. Match the following numbers of species with the best description.
   __ 750 thousand
   __ 1.5 million
   __ 3 million
   __ 30 million
   __ 100 million
   a. current number of known insects
   b. Erwin's estimate of the total number of arthropod species
   c. lower estimate of the total number of all species
   d. upper estimate of the total number of all species
   e. current number of total known species

3. It has taken biologists more than two centuries to catalog the ________ known species. Yet many believe we have barely scratched the surface and that ________ percent or more of all living species remain to be identified.
   a. 750 thousand; 90
   b. 1.5 million; 90
   c. 3 million; 50
   d. 30 million; 50
   e. 100 million; 10

4. Despite greatly increased efforts by biologists to reach remote areas such as forest canopies and ocean depths, many of the creatures that live in these have yet to be seen in their natural habitats. (True or False)

5. Which of the following statements about mass extinctions is *false*?
   a. Research indicates that we are on the verge of a new mass extinction event.
   b. Only 73 percent of the large mammal species that roamed earth just 10,000 years ago exist today.
   c. Current extinction rates are higher than in any previous mass extinction event.
   d. The greatest loss of species is currently occurring in tropical rainforests.
   e. Human activity is the main cause of the new ongoing mass extinction.

6. The fossil record indicates that humans have been driving species into extinction for a very long time. In particular, the evidence for human influence on the extinction of large animal species such as saber-toothed cats, mammoths, and flightless birds goes back at least ________ years.
   a. 1,000
   b. 5,000
   c. 10,000
   d. 30,000
   e. 100,000

7. Scientists can very easily and accurately determine the number and rate of species being lost at the present time. (True or False)

8. Which of the following is *not* believed to be a threat to biodiversity?
   a. "fogging" research conducted on the rainforest canopy
   b. suburban development
   c. pollution
   d. introduction of foreign species
   e. climate change

9. The declines in species numbers for both Lake Victoria cichlid fish and native birds in Hawaii
   a. are clearly the result of changes in climate.
   b. may be the result of decreased human activity.
   c. are mainly the result of introduced foreign species.
   d. are caused by unknown factors.
   e. none of the above

10. The single most important way in which humans can help stem the tide of species loss is for them to
    a. learn to control the climate.
    b. reduce the growth of human populations.
    c. decrease the amount of ultraviolet light.
    d. increase the introduction of foreign species into new habitats.
    e. control the release of toxic pollutants.

11. Not counting the present high rate of extinctions, there have been a total of 50 mass extinctions during the history of life on Earth. (True or False)

12. Which of the following has *not* been hypothesized to be the cause of a mass extinction on earth?
    a. climate change
    b. increased volcanic activity
    c. asteroids
    d. human activity
    e. none of the above

13. The ocean reefs have recovered from multiple mass extinction events. In each case, the recovery was rapid, resulting in even greater biodiversity than before. (True or False)

14. Maintaining biodiversity is a primary key to a healthy ecosystem. When ecosystems are healthy, humans can harvest a wide variety of useful materials. Which of the following useful products comes from nonhuman species?
    a. one-fourth of all prescription drugs
    b. the oxygen that we breathe
    c. meat and dairy products
    d. vegetables and wood
    e. all of the above

Conceptual Understanding

1. The sudden disappearance of the golden toad from the high tropical forests of Costa Rica should give us cause for alarm because
    a. until recently this species was very abundant.
    b. amphibian species all over the globe are declining at the same time.
    c. increased exposure to ultraviolet light, partly caused by humans, might be to blame.
    d. pollution and introduced diseases are possible culprits.
    e. all of the above

2. In terms of species currently known to science, the number of plant species is greater than the number of animal species. (True or False)

3. In the species-scape pictured in Figure A.4 of your textbook, the proper size for an ant would be comparable to the size shown for the
    a. elephant.
    b. annelids.
    c. sponges.
    d. dragonfly.
    e. plants.

4. Biologists often rely on estimates of species numbers since it is impossible to get a truly accurate count. By definition, estimates are based on assumptions about the accuracy of measurements and how representative the samples are that are used to generate the estimates. Which of the following is an assumption upon which Terry Erwin's estimate of total arthropod species number is based?
    a. There are 50,000 tropical tree species.
    b. Biodegradable chemical fogger kills all insects equally.
    c. Roughly 40 percent of all arthropod species are beetles.
    d. Canopies have twice the number of insects found elsewhere in tropical forests.
    e. all of the above

5. Although there have been many possible causes identified for the disappearance of species, foremost among the current threats to biodiversity is the
    a. threat of a collision with a large asteroid.
    b. effect of climate change.
    c. destruction of habitat by humans.
    d. damage done by volcanic eruptions.
    e. none of the above

6. Which of the following statements about worldwide changes in species numbers is *false*?
    a. Twenty percent of total bird species have disappeared in the past 2000 years.
    b. Human population growth seems to have little effect on other species.
    c. 73 percent of large mammal species have disappeared over the last 10,000 years.
    d. Overall species loss from rainforests is occurring at a rate of nearly 27,000 per year.
    e. Freshwater fish species have been reduced by 20 percent in recent history.

7. Match each term with the best description.
    __ foreign species
    __ mass extinction
    __ ecotron
    __ productivity
    __ habitat
    a. small experimental ecosystem
    b. actual mass of plant matter in a habitat
    c. disappearance of many groups at one time
    d. not native to the habitat being observed
    e. physical and biological place where species live

8. Introducing non-native species into foreign habitats increases overall species diversity and rarely has negative side effects. (True or False)

9. The recent dramatic decline of Alpine androsace described in your textbook is likely due to
    a. the introduction of foreign species.
    b. global warming.

c. suburban development.
d. water pollution.
e. forces yet to be identified.

10. As mammals, humans ultimately owe a debt of gratitude to the mass extinction that occurred _______ million years ago, which gave the mammals the opportunity to flourish.
a. 65
b. 206
c. 250
d. 350
e. 440

11. There have been several mass extinctions during the history of life on Earth. However, unlike the possible mass extinction event currently underway, previous ones _______ caused by humans. The fossil record indicates that species numbers recovered from mass extinctions only after _______ of years.
a. were; many thousands
b. were not; many thousands
c. were; many millions
d. were not; many millions
e. none of the above

12. Despite widescale efforts, recycling in the United States has had little impact on the amount of waste produced by consumers. (True or False)

13. What do a capybara, green iguana, Pacific yew tree, and pineapple all have to do with the issue of whether humans should work to maintain biodiversity?
a. Each is regarded as a key species in its environment.
b. They are all plants, and plants are the key to high productivity.
c. They all provide important products that are used by people.
d. These species are known to have survived previous mass extinctions.
e. They are all the source of important medicines.

14. According to the textbook, a habitat that has _______ species of plants will be _______ productive than a habitat with _______ species of plants.
a. more; less; more
b. fewer; less; more
c. fewer; more; more
d. more; more; fewer
e. One cannot answer this question with the information provided.

Related Activities

- Find three different sites on the World Wide Web that address the issue of biodiversity. Write a one-page essay that summarizes how the concept of biodiversity is relevant to the organization maintaining the Web page, and compare the approach each takes to support (or refute!) the need for biodiversity.
- Visit your school library and locate a research article that gives information on the total number of species present in an oceanic environment. Sketch a "speciesscape" for this marine ecosystem (refer to the example in Figure A.4 of your textbook). Remember to draw the size of your representative organisms in proportion to their percentage of the total species count.
- Review the Biology Matters section of the interlude, which focuses on the amount of waste produced by the average college student. Visit your local student services department on campus and obtain information on the recycling programs currently in place on your campus. Compose a one-page summary of how aggressive your campus is in promoting recycling programs. At the end of your summary, provide a list of suggestions you can think of to improve your campus's recycling efforts.

Answers and Explanations

Factual Knowledge

1. **c** Review the information on the number of known species illustrated in Figure A.3 of your textbook.

2. **a** 750 thousand
e 1.5 million
c 3 million
b 30 million
d 100 million

3. **b** The current species count stands at about 1.5 million, but given the estimates from work such as that of Terry Erwin in the tropical rainforest canopy, it appears that biologists have barely scratched the surface of species identification and that upwards of 90 percent of all species remain to be classified.

4. **True** Based on what little sampling has been done, remote habitats probably will yield large numbers of new species, as well as species that are already known but have never actually been seen in their natural habitat.

5. **b** Recall from the text that 73 percent of the Earth's large mammal species that existed 10,000 years ago are now extinct.

6. **c** The arrival of humans 10,000 years ago in places like Australia and North America coincided with the rapid decline of many large animal and bird species that would have been easy prey.

7. **False** The total number of species on Earth is currently unknown. Therefore, it would be impossible to know with certainty the exact number of species currently being lost and the rate at which they are disappearing.

8. **a** Scientific research, carefully conducted, poses no threat to biodiversity. Rather, the understanding that develops from such research serves to help preserve biodiversity.

9. **c** Introduced species are a big problem for biodiversity. Lake Victoria's natural diversity of cichlid fish has been greatly reduced by introduction of the Nile perch, and many of Hawaii's native birds are losing out to predation by feral cats, mongooses, and other nonnative animals.

10. **b** Increasing the rate at which nonnative species are introduced would only make the problem worse. All the other choices would help promote biodiversity, but clearly the most important contribution of humans is to control the growth of our own species.

11. **False** The fossil record indicates five, not 50, previous mass extinctions.

12. **e** All of the factors listed have been hypothesized to be responsible for mass extinctions.

13. **False** Recall from the text that while the recovery from mass extinction events, particularly in the ocean reefs, did occur, the time required for each recovery was on the order of 5 to 10 million years.

14. **e** All these are products that come from nonhuman species. Therefore, the more diverse an ecosystem, the more usable products may be derived from that system.

Conceptual Understanding

1. **e** It is believed there is no single cause for the disappearance of the golden toad. All of the suggested causes have likely contributed to the loss of the golden toad.

2. **False** Recall that the animal group would include both the insects and the non-insect animals. Refer to Figure A.3 for this information.

3. **d** Since ants are part of the insect group, in the species-scape they should be the same size as other insects, like the dragonfly.

4. **e** Assumptions are inherent in much of what goes into making estimates, from the chemical effects of foggers to the mathematical variables that go into calculating the actual estimates.

5. **c** Although each of the other factors listed can adversely influence biodiversity, nothing has a greater impact on species numbers than the deterioration or destruction of habitats resulting from human activity.

6. **b** Just the opposite of this statement is true. Human population growth has a major impact on other species.

7. **d** foreign species
c mass extinction
a ecotron
b productivity
e habitat

8. **False** The introduction of non-native species often has unintended and negative environmental consequences, which ultimately result in a loss of biodiversity.

9. **b** The temperature increase from global warming is forcing these plants, which have low heat tolerances, farther and farther up the mountains. Their habitat is thus becoming restricted, and as a result their numbers are dwindling.

10. **a** The mass extinction of 65 million years ago triggered the disappearance of the dinosaurs, the dominant land vertebrates of the time. This opened the door for the mammals to diversify and become the next lineage of vertebrates to dominate the terrestrial world. Humans eventually evolved from the mammalian subgroup known as primates.

11. **d** Humans had not even evolved at the time of the most recent mass extinction 65 million years ago. The fossil record indicates that while species numbers do rebound after mass extinctions, the time course for this recovery is many millions of years.

12. **False** Recall from the Biology Matters section of the Interlude that the United States Environmental Protection Agency reports that up to 42 percent of all paper used in this country is now recycled and that due to recycling, 64 million tons of waste were diverted from landfills and incinerators. The United States currently recycles 28 percent of its total waste.

13. **c** The best answer is that all four are examples of species that produce valuable products for human use. Although all are important, none is considered a key species in its ecosystem. Two are animals, and the other two are plants. Some yield medicines, while the others are used mainly for food. Some, but not all, were present at the time of the last mass extinction.

14. **d** Habitats tend to be more productive, to produce more plant matter, as the biodiversity increases. They also tend to be more resilient and resistant to disease.

CHAPTER 4 Chemical Building Blocks

Getting Started

Below are a few questions to consider prior to reading Chapter 4. These questions will help guide your exploration and assist you in identifying some of the key concepts presented in this chapter.

1. What do scientists mean when they refer to "blueberries" on the surface of Mars?
2. What distinguishes an element's atomic number from its atomic mass number?
3. What type of chemical bond results from the time-sharing of electrons between two atoms?
4. What are the differences between a solution, a solute, and a solvent?
5. Why are nonpolar molecules considered to be hydrophobic?
6. What biological role do buffers play in regulating the pH of a solution?
7. Aside from a type of food you might purchase at your local grocery store, what does the term *organic* mean?
8. Walking on water is no small task: however, the water-strider does it with ease. How does this small insect accomplish this feat?
9. Consumption of too many saturated fats can result in a variety of diseases. What is the primary difference between saturated and unsaturated fats?

A Guide to the Reading

The following concepts typically give students the most difficulty when exploring the content in Chapter 4 for the first time. For each concept, one or more references have been identified that may help you gain a better understanding of these potentially problematic areas.

Radioisotopes

Each element is characterized by its atomic number, which is the number of protons found in the atom's nucleus. Recall that the nucleus also contains a certain number of neutrons as well. The number of neutrons usually equals the number of protons for a particular element, although the number of neutrons may vary. An isotope is an element with the same number of protons, but a different number of neutrons. The example described in the chapter is carbon, which can exist as C-12 (6 protons + 6 neutrons) or C-14 (6 protons + 8 neutrons). Isotopes such as C-14 may in fact be unstable as a result of having more than the normal number of neutrons. Unstable isotopes tend to give off energy (in the form of radiation) through the process of radioactive decay. As a result, these unstable isotopes are often referred to as radioisotopes. High energy radioisotopes can cause damage to DNA and result in cancer. Lower energy radioisotopes can be useful for medical diagnostics as they can be used to trace molecules as they proceed through a biological process, serving as a chemical "tracer."

For more information on this concept, be sure to focus on:

- In Section 4.1, Some elements can exist as isotopes with different atomic mass numbers
- Figure 4.2, Isotopes in Medicine

Nonpolar Molecules

As discussed in the chapter, molecules can be held together through the process of sharing electrons. When electrons are not shared equally between bonded atoms, this results in the uneven distribution of electrical charge in the resulting molecule—one half of the molecule will have a slight positive charge; the other half will carry a slight negative charge. Such molecules are called *polar*. A good example of a polar molecule would be water. Polar molecules tend to be attracted to one another through these partial electrical charges. The electrical interactions that hold these molecules together are called *hydrogen bonds*. Polar molecules as a class will be soluble in water since they can interact with the partial electrical charges of the water molecules. Molecules that have an affinity for interacting with water like this are called *hydrophilic*, or "water-loving." In contrast, molecules that are held together by the equal sharing of electrons tend to have no partial electrical charge and are called *nonpolar*. With no charges available to interact with partially charged water molecules, nonpolar molecules tend to be insoluble in water. A good example of a nonpolar molecule would be a fat molecule, or lipid. These molecules are called *hydrophobic*, or "water-fearing."

For more information on this concept, be sure to focus on:

- In Section 4.3, Hydrogen bonds are important temporary bonds
- Figure 4.4, Hydrogen Bonds Determine the Properties of Water and How Other Compounds Interact with Water

pH Scale

The acidity of a solution is a measure of the solution's concentration of free hydrogen ions. The higher the concentration of free hydrogen ions in a solution, the more acidic that solution will be. The lower the concentration of free hydrogen ions, the less acidic (and more basic) the solution will be. The pH scale was developed to describe how acidic a solution is by measuring the concentration of free hydrogen ions in a solution. The scale ranges from 0 to 14, with a lower pH value indicating a higher concentration of hydrogen ions in solution and a more acidic solution. At the other end of the scale, higher pH values indicate a lower concentration of hydrogen ions and a more basic solution. Most biological systems maintain a pH around 7, which is considered neutral. Maintaining proper pH is vital to the survival of biological systems. To assist in this process, special compounds called *buffers* are utilized that help to absorb excess hydrogen ions when a solution becomes too acidic or donate hydrogen ions when a solution becomes too basic. The key to understanding the relationship of pH and acidity is realizing that the pH scale is actually inverted in terms of hydrogen ion concentration (i.e., low pH = high acidity; high pH = low acidity).

For more information on this concept, be sure to focus on:

- In Section 4.4, Acids and bases affect the pH of water
- In Section 4.4, Buffers prevent large changes in pH
- Figure 4.5, The pH Scale

Saturated vs. Unsaturated Fats

Fatty acids are molecules that typically consist of long chains of carbon and hydrogen atoms. These long chains can take one of two forms. When each carbon atom in a chain is attached to carbon atoms on either side and two hydrogen atoms, the chain is said to be "saturated" or "full" of hydrogen atoms. Saturated fatty acid chains tend to be straight, which enables them to pack together tightly. Fat molecules consisting of mostly saturated fatty acids tend to be solid at room temperature. In contrast, fatty acid chains in which two or more of the carbon atoms are held together with double covalent bonds will have fewer than the maximum number of two hydrogen atoms per carbon. As a result, unsaturated fatty acid chains will have small bends, or "kinks," in their chain. These kinks tend to prohibit the molecules from packing together tightly. Fats consisting of mostly unsaturated fatty acids will tend to be liquid at room temperature. The key to understanding the difference between saturated and unsaturated fatty acids is understanding that saturated fats have carbon chains that are full or "saturated" with hydrogen atoms while the carbon chains of unsaturated fats will have fewer than the maximum.

For more information on this concept, be sure to focus on:

- In Section 4.5, Fatty acids store energy and form membranes
- Figure 4.14, Saturated and Unsaturated Fatty Acids

Tying It All Together

Several concepts presented in this chapter may be revisited and discussed in greater detail in subsequent chapters, including:

Biological Membranes

- Chapter 5 – Section 5.2, The Plasma Membrane: Separating Cells from the Environment

Energy

- Chapter 7 – Section 7.1, The Role of Energy in Living Systems

Macromolecules and Nutrition

- Chapter 7 – Section 7.2, Using Energy from the Controlled Combustion of Food

Chemical Reactions

- Chapter 7 – in Section 7.2, Chemical reactions are governed by simple transformations of energy
- Chapter 7 – Section 7.3, How Cells Speed Up Chemical Reactions

Radiation in Science

- Chapter 12 – in Section 12.1, The genetic instructions of a virus are contained in its DNA

DNA

- Chapter 12 – Section 12.2, The Three-Dimensional Structure of DNA

Building Biological Molecules

- Chapter 12 – Section 12.3, How DNA Is Replicated
- Chapter 13 – Section 13.5, Translation: Information Flow from mRNA to Protein

Practice Questions

Factual Knowledge

1. The elements hydrogen, nitrogen, and carbon
   a. can all form ionic bonds with other elements.
   b. contain neutrons in their atomic nuclei.
   c. have numerous electrons that orbit the inner core of their atoms.
   d. are common elements in the molecules that make up living organisms.
   e. all of the above

2. If two electrons come close to one another, they
   a. are attracted by their opposite electric charges.
   b. are repelled by their like electric charges.
   c. form an element.
   d. form a hydrogen bond.
   e. produce an ion.

3. Which of the following statements about covalently bonded molecules is *false*?
   a. Electrons are shared between the atoms that make up the molecule.
   b. Noncovalent bonds may also be present, especially if the molecule is large.
   c. Such molecules often result from the interaction of ionized atoms.
   d. The chemical bonds that hold the molecules together are relatively strong.
   e. Protons and electrons are present in the atoms that make up the molecule.

4. $H_{22}O$ and CH_4 are both examples of
   a. molecules.
   b. ions.
   c. elements.
   d. acids.
   e. all of the above

5. Which of the following statements about water is *true*?
   a. It contains two of the most common elements on Earth.
   b. It is critical for many of the chemical processes found in living systems.
   c. Polar substances that can form hydrogen bonds will dissolve in it.
   d. Nonpolar substances like oils are not soluble in it.
   e. all of the above

6. Which of the following would *not* be considered an isotope?
   a. ^{2}H (Deuterium)
   b. ^{3}H (Tritium)
   c. ^{12}C
   d. ^{14}C
   e. ^{32}P

7. Hydrogen bonds and hydrophilic interactions are
   a. weak chemical bonds that hold together atoms within a molecule.
   b. strong chemical bonds that hold together atoms within a molecule.
   c. weak chemical bonds that link separate molecules together.
   d. strong chemical bonds that link separate molecules together.
   e. interactions that hold protons and neutrons together within the atomic nucleus.

8. The main difference between an acid and a base is that
   a. bases are polar molecules and acids are not.
   b. acids are polar molecules and bases are not.
   c. bases donate hydrogen ions in water while acids accept hydrogen ions.
   d. acids donate hydrogen ions in water while bases accept hydrogen ions.
   e. one can act as a buffer while the other cannot.

9. A pH value of 0 represents the lowest possible concentration of free hydrogen ions in a solution. (True or False)

10. The role of chemical buffers in living systems is to maintain the hydrogen ion concentration within narrow limits favorable for the internal environment of cells. (True or False)

11. A solution with pH = 5 is _______ than a solution with pH = 7.
    a. 2 times more basic
    b. 10 times more basic
    c. 2 times more acidic
    d. 10 times more acidic
    e. 100 times more acidic

12. The molecule ATP is most closely related to which class of macromolecules?
    a. nucleic acids
    b. fats
    c. carbohydrates
    d. proteins
    e. membranes

13. Which of the following is a polymer of amino acids?
    a. ATP
    b. fats
    c. carbohydrates
    d. proteins
    e. membranes

14. In a saturated fat, you would expect to find all of the following except
    a. single-bonded carbon atoms.
    b. double-bonded carbon atoms.
    c. hydrogen atoms.
    d. nonpolar bonds.
    e. covalent bonds.

15. Match each term with the best description.
    __ phospholipid
    __ carbohydrate
    __ hydrophobic
    __ acidity
    __ ionic bond
    a. major component of biological membranes
    b. a nonpolar compound's interaction with water
    c. an attraction of oppositely charged atoms
    d. the cause of tartness in lemon juice
    e. a polysaccharide

Conceptual Understanding

1. In which of the following molecules would you expect to find electrons shared between the atoms that make up the molecule?
    a. water
    b. hydrogen gas
    c. methane
    d. hydrogen sulfide
    e. all of the above

2. The strongest chemical linkages are found in molecules in which atoms are held together by _______ bonds.
    a. noncovalent
    b. covalent
    c. ionic
    d. hydrogen
    e. hydrophilic

3. A compound capable of forming hydrogen bonds with water
    a. is probably held together by noncovalent bonds.
    b. contains at least some polar covalent bonds.
    c. should act as a good buffer for acids and bases.
    d. does not dissolve well in water.
    e. is hydrophobic.

4. Salt and sugar both dissolve well in water, but for different reasons. In the case of salt, water molecules
    a. form ionic bonds with the positively and negatively charged ions.
    b. form nonpolar covalent bonds with the positively charged ions only.
    c. surround the separated ions because of their charge.
    d. share electrons with the ions to make polar covalent bonds.
    e. none of the above

5. Chemical reactions always begin and end with the same number and type of atoms, although those atoms are usually rearranged during the reaction. (True or False)

6. What will happen if you add a mild acid to a solution that is buffered at pH = 7?
    a. The pH will drop well below 7.
    b. The pH will rise well above 7.
    c. The pH will not change much.
    d. Ionic bonds will form between the acid and the buffer.
    e. Covalent bonds will form between the acid and the buffer.

7. Which of the following is *not* an example of a compound that would fall into the category of monomers?
    a. monosaccharides
    b. DNA
    c. amino acids
    d. nucleotides
    e. none of the above

8. Proteins, DNA, and starch are all
   a. carbon-based compounds.
   b. polymers.
   c. held together mainly by covalent bonds.
   d. made of atoms that contain protons and electrons.
   e. all of the above

9. A biological polymer is known to contain alanine, tyrosine, and lysine. This polymer is
   a. DNA.
   b. a strong base.
   c. a phospholipid.
   d. a protein.
   e. held together by hydrogen bonds.

10. Glucose is to cellulose as glycerol is to
    a. DNA.
    b. starch.
    c. protein.
    d. phospholipids.
    e. fatty acids.

11. Consider Figure 4.11 in your textbook. Assume that a small protein is made exclusively from hydrophobic amino acids. This protein would be considered
    a. polar.
    b. nonpolar.
    c. neutral.
    d. a buffer.
    e. none of the above

12. Which of the following components would *not* be found in a typical phospholipid bilayer?
    a. cholesterol
    b. glycerol
    c. fatty acids
    d. guanine
    e. carbohydrates

13. In a phospholipid bilayer, the hydrophobic tails of the phospholipid molecules are pointed outward so that they are in maximum contact with water. (True or False)

Related Activities

- Refer to the discussion in the opening section of the chapter of the discovery of "blueberries" and jarosite on the surface of a weathered rock from Mars. Explain why this finding has been so exciting to scientists.
- Make a list of foods that you ate today and identify the basic classes of molecules in each. Are all of the classes represented? How does this nutritional analysis relate to the concepts of biological chemistry discussed in the chapter?
- Search the World Wide Web for information related to efforts to engineer heat-resistant proteins as described in the Proteins That Can Take the Heat section at the end of the chapter. Describe some of the practical applications of this research to everyday life.

Answers and Explanations

Factual Knowledge

1. **d** Recall that hydrogen does not typically contain neutrons or have multiple electrons orbiting the nucleus. Ionic bonds are also formed between elements that can easily donate or accept electrons. All three of these elements are common to living organisms.
2. **b** Electrons have a negative electric charge and therefore would repel each other.
3. **c** Ionic bonds are caused by the attraction of oppositely charged ions. Covalent bonds result from the sharing of electrons between atoms.
4. **a** Fixed ratios of specific atoms yield molecules, such as water and methane.
5. **e** Water is made of hydrogen and oxygen, which are abundant elements. It is also a key component of many biological processes. Since water is polar, it will form hydrogen bonds with other polar molecules but not with nonpolar compounds.
6. **c** Recall that the atomic mass number of carbon is 12. Therefore, C-12 would not be considered an isotope while C-14 would.
7. **c** These bonds form between molecules, or within specific parts of molecules, to help stabilize structure.
8. **d** Acids donate hydrogen ions to lower the pH of a solution, while bases accept hydrogen ions to raise the pH.
9. **False** A pH of 0 represents the highest possible concentration of hydrogen ions.
10. **True** Buffers are substances that can donate or accept hydrogen ions as necessary to maintain a stable pH.

11. **e** The pH scale is logarithmic, which means that each step on the scale represents a 10-fold increase or decrease in hydrogen ion concentration. A pH of 7 is neutral, and a pH of 5 is mildly acidic.

12. **a** The "A" in ATP stands for adenosine, a modified adenine nucleotide.

13. **d** Polymers are repeating units of monomers. Amino acids are the monomers used to make proteins.

14. **b** "Saturated" means that the maximum number of hydrogen atoms are attached to each carbon, precluding the possibility of double bonds between carbon atoms.

15. **a** phospholipid
    e carbohydrate
    b hydrophobic
    d acidity
    c ionic bond

Conceptual Understanding

1. **e** All of these are covalently bonded molecules. Electrons are thus shared between the atoms within the molecule.

2. **b** Because electrons are actually shared between atoms, covalent bonds are the strongest types of chemical bonds.

3. **b** Hydrogen bonds form between water molecules and the charged portions of other polar covalent molecules.

4. **c** The electrical charge present in salt ions attracts water molecules, which then surround them, keeping them separate from their electrically charged counterparts.

5. **True** Although new products are made in chemical reactions, the equation must always balance in terms of number and type of atoms involved. That is, matter can be neither created nor destroyed, only rearranged.

6. **c** Recall that the function of a buffer is to maintain the pH within a narrow range.

7. **b** DNA is a polymer made from nucleotide monomers.

8. **e** All of these statements are true.

9. **d** Alanine, tyrosine, and lysine are all amino acids used to make proteins.

10. **d** Glucose is a building block used to make cellulose. Glycerol is a building block used to make phospholipids.

11. **b** Recall that hydrophobic amino acids are also nonpolar. As a result, this protein would be repelled by water.

12. **d** Recall that the nucleic acid guanine is one of the building blocks of DNA. You would not typically find guanine associated with a phospholipid bilayer.

13. **False** Just the opposite: The tails are pointed inward to avoid water. See the diagram in Figure 4.16 in your textbook.

CHAPTER 5 Cell Structure and Compartments

Getting Started

Below are a few questions to consider prior to reading Chapter 5. These questions will help guide your exploration and assist you in identifying some of the key concepts presented in this chapter.

1. What provides the propulsive force that pushes the bacterium *Listeria* as it travels through an infected cell?
2. Who was Robert Hooke, and what was he the first to observe?
3. What is the fluid mosaic model?
4. What are the "little organs" of the cell called?
5. If the cell were considered as a tiny factory, which structure would serve as the shipping department? Which structure would serve as the power plant?
6. What are the three main components of the cytoskeleton, and what are their functions?
7. In a cell that is capable of swimming through water, what tiny structures would be analogous to the oars of a rowboat?

A Guide to the Reading

The following concepts typically give students the most difficulty when exploring the content in Chapter 5 for the first time. For each concept, one or more references have been identified that may help you gain a better understanding of these potentially problematic areas.

Fluid Mosaic Model

The plasma membrane serves as the outer boundary of the cell. As discussed in Chapter 4, membranes are composed primarily of phospholipids arranged in a bilayer. Unfortunately, this bilayer arrangement restricts the movement of many molecules across the membrane. In order for the cell to import nutrients, export wastes, and communicate with its surroundings through the use of signaling molecules, it employs the services of special membrane proteins that are embedded in the phospholipid bilayer. These proteins perform specialized functions, such as the transport of larger molecules across the membrane. It is important to note that the proteins embedded in the plasma membrane are typically free to move laterally within the plane of the membrane. Since the membrane is composed of a mosaic of protein and phospholipids and the proteins are free to move fluidly, the plasma membrane of cells is often referred to as a *fluid mosaic.* This characteristic is important for the cell's ability to move and to communicate with its environment.

For more information on this concept, be sure to focus on:

- Section 5.2, The Plasma Membrane: Separating Cells from Their Environment
- Figure 5.1, Proteins Are Embedded in the Plasma Membrane

Nuclear Envelope

As discussed in the chapter, the nucleus serves as the "administrative center" of the cell and is the central repository for genetic information in eukaryotes. It is a relatively large, distinct structure that houses the cell's DNA. In order to keep nuclear contents and the cytosol separate, the nucleus is enclosed by two membranes—referred to as the nuclear envelope. The nuclear envelope contains numerous holes, or "pores," which provide a passageway for particular molecules to enter or exit the nucleus. These nuclear pores are very similar to the channels that appear within the plasma membrane, regulating the passage of larger molecules into or out of the cell.

For more information on this concept, be sure to focus on:

- In Section 5.4, The nucleus is the repository for genetic material
- Figure 5.3, The Nucleus

Movement Between Compartments

The cytosol of the eukaryotic cell contains numerous organelles that work together to carry out the activities of the cell. Part of this includes the production and export of proteins. This process begins at the "factory floor" of the cell, the endoplasmic reticulum. Tiny structures called *ribosomes* attached to the surface of the endoplasmic reticulum function to produce proteins directly inside the membrane-enclosed ER (the lumen). In order to complete their journey, these proteins must escape the confines of the ER for processing and sorting in a different organelle—the Golgi apparatus. But how do they get there? The answer is through the use of tiny, membrane-bound transport structures called *vesicles*. Vesicles are produced when parts of the ER begin to "bud" off of the organelle. Contents of the lumen of the ER, including the proteins destined for the Golgi apparatus, become trapped inside these buds, which finally break off, forming a transport vesicle. The transport vesicle then makes its way through the cytosol until it fuses with a nearby Golgi apparatus, dumping the trapped contents (including the protein) inside the Golgi. Once the proteins are processed inside the Golgi, they can then be transported by another vesicle to their final destination. The budding and fusion of transport vesicles is made possible by the fluid nature of the bounding membranes. Since the inner portion of the phospholipid bilayers is hydrophobic, they will have a tendency to fuse together, just like two bubbles floating in the air or two droplets of oil floating on the surface of water might fuse together.

For more information on this concept, be sure to focus on:

- In Section 5.4, The endoplasmic reticulum is the site of manufacture for lipids and many proteins
- Figure 5.5, How Vesicles Move Proteins and Lipids from One Compartment to Another

Cell Motility

The cytoskeleton of the cell serves to provide both structure and movement to a cell. It consists of three distinct types of filaments: microtubules, intermediate filaments, and microfilaments. Microtubules are the largest of these structures. In eukaryotes, they facilitate cell motility in two ways. The first way involves the creation of a tiny system of "tracks," analogous to a subway system, starting at the center of the cell (adjacent to the nucleus) and branching outward toward the plasma membrane. These tiny tracks are used by structures such as transport vesicles to crawl along to reach their final destination. Microtubules also provide the rigidity and flexibility necessary to produce tiny rowing (cilia) or beating (flagella) structures that serve to propel cells through a fluid medium.

In contrast, microfilaments are the smallest of the cytoskeletal structures. In eukaryotes, they facilitate cell movement by constantly adjusting their length and rearranging their alignment at the cell surface. The growth and realignment of the microfilaments causes small protrusions, termed *pseudopodia*, to emerge from the cell, allowing it to grab hold and crawl along a surface. Intermediate filaments are not believed to serve a motility function in cells, but rather provide the cell with structural support.

For more information on this concept, be sure to focus on:

- In Section 5.5, Microtubules support movement inside the cell
- In Section 5.5, Microfilaments are involved in cell movement
- In Section 5.5, Cilia and flagella act like oars and propellers
- Figure 5.11, Kinds of Filaments in the Cytoskeleton
- Figure 5.12, Microfilaments Allow Cell Movement
- Figure 5.13, Eukaryotic Cilia and Flagella

Tying It All Together

Several concepts presented in this chapter build upon concepts presented in previous chapters and are also revisited and discussed in greater detail in subsequent chapters, including:

Prokaryotes vs. Eukaryotes

- Chapter 3 – in Section 3.1, Organisms can also be identified as prokaryotes and eukaryotes

Atomic Composition of Life

- Chapter 4 – Section 4.1, Atoms Make Up the Physical World

Biological Membranes

- Chapter 4 – in Section 4.5, Fatty acids store energy and form membranes

Cell Signaling and Communication

- Chapter 6 – Section 6.6, Signaling Molecules in Cell Communication

Energy in Living Systems

- Chapter 7 – Section 7.1, The Role of Energy in Living Systems
- Chapter 8 – Section 8.3, Catabolism: Breaking Down Molecules for Energy

Photosynthesis

- Chapter 8 – Section 8.2, Photosynthesis: Capturing Energy from Sunlight

DNA

- Chapter 12 – Section 12.1, The Search for the Genetic Material

Protein Synthesis

- Chapter 13 – Section 13.5, Translation: Information Flow from mRNA to Protein

Practice Questions

1. In terms of basic structure, elephant cells and oak tree cells both
   a. are prokaryotes.
   b. have chloroplast.
   c. have a cell wall.
   d. have mitochondria.
   e. all of the above

2. The simplest way to distinguish a prokaryotic from a eukaryotic cell is to check for
   a. a plasma membrane.
   b. a nucleus.
   c. DNA.
   d. proteins.
   e. whether the cell is a single-celled organism.

3 In which of the following cell types would you see more cellular detail using an electron microscope as compared to a light microscope?
   a. animal
   b. plant
   c. bacterial
   d. protist
   e. all of the above

4. Which of the following structures would you expect to find in a bacterium?
   a. nucleus
   b. plasma membrane
   c. Golgi apparatus
   d. lysosome
   e. endoplasmic reticulum

5. Which organelle is surrounded by a double phospholipid bilayer with many large pores?
   a. nuclear envelope
   b. plasma membrane
   c. Golgi apparatus
   d. mitochondrion
   e. all of the above

6. Which of the following statements about prokaryotes is *false*?
   a. Just like eukaryotes, prokaryotes have a plasma membrane.
   b. Prokaryotic cells concentrate important materials for the cell's survival.
   c. The substance known as *cytosol* is found within the bacterial nucleus.
   d. Prokaryotic cells are much smaller than most eukaryotic cells.
   e. Ancient prokaryotes may have given rise to some eukaryotic organelles.

7. Which of the following structures would be considered an organelle?
   a. microtubule
   b. cytosol
   c. vacuole
   d. plasma membrane
   e. DNA

8. Proteins that are destined for insertion into a membrane are one product of the rough ER. (True or False)

9. The Golgi apparatus is one destination for substances that have been packaged into vesicles by the smooth ER. (True or False)

10. Lysosomes are specialized vesicles in ________ that contain digestive enzymes for the breakdown of food. A related organelle known as a *vacuole*, which is found in ________, also contains enzymes but in addition may act as a storage organelle for nutrients or water.
    a. animals; plants and fungi
    b. plants; animals and fungi
    c. fungi; animals and plants
    d. plants and fungi; animals
    e. animals and plants; fungi

11. Both chloroplasts and mitochondria
    a. have multiple membranes.
    b. have highly structured innermost membranes.

c. are found only in eukaryotic cells.
d. are involved in cellular energy processing.
e. all of the above

12. In which of the following cell organelles would you expect to find the biochemical reactions that harness energy from the breakdown of sugar molecules to synthesize large amounts of ATP?
a. lysosome
b. vesicles
c. chloroplast
d. mitochondrion
e. plasma membrane

13. Microtubules, motor proteins, and actin filaments are all part of the
a. mechanism of photosynthesis that occurs in chloroplasts.
b. rough ER in prokaryotic cells.
c. cytoskeleton of eukaryotic cells.
d. process that moves small molecules across cell membranes.
e. none of the above

14. You are asked to examine a cell using a powerful light microscope. The image you see has a clearly defined nucleus and mitochondria. It also has a large vacuole and mitochondria. From what group of organisms did this cell most likely come?
a. Bacteria
b. Fungi
c. Plants
d. Animals
e. cannot tell

15. Which of the following macromolecules are found in the plasma membrane?
a. lipids only
b. lipids and proteins
c. lipids, proteins, and nucleic acids
d. proteins and nucleic acids
e. proteins only

16. According to the fluid mosaic model of cell membranes
a. proteins are rigidly fixed in the phospholipid bilayer.
b. the most common types of molecules in the membrane are proteins.
c. basic membrane structure results from the ways in which proteins interact with water.
d. the membrane is a highly mobile mixture of phospholipids and proteins.
e. the unique properties of cell types are determined by their phospholipids.

17. The best way to identify a cell as either prokaryotic or eukaryotic is to determine whether
a. it came from a single-celled or multicellular organism.
b. it contains DNA enclosed within a membrane-bound compartment.
c. it has a cell wall.
d. it has cytosol.
e. all of the above

18. Eukaryotic cells are more efficient than prokaryotes because their internal compartmentalization
a. makes each compartment nutritionally independent of all others.
b. allows for specialization through the subdivision of particular tasks.
c. permits the unregulated flow of materials around the cell.
d. eliminates the need for communication with the external environment.
e. reduces overall cell size.

19. In a microscope, you observe an organelle that consists of a series of interconnected tubes and appears to be connected to the outer membrane of the nucleus in a eukaryotic cell. What is the likely identity of this organelle?
a. mitochondrion
b. lysosome
c. Golgi apparatus
d. vacuole
e. endoplasmic reticulum

20. Consider a protein that is destined to be embedded in the plasma membrane. This protein must first be assembled in the ________ and then transported to the ________ for processing and sorting.
a. smooth ER; rough ER
b. mitochondria; Golgi apparatus
c. Golgi apparatus; transport vesicle
d. rough ER; Golgi apparatus
e. lysosome; smooth ER

21. The vesicles that bud off of the ER contain free-floating molecules from the lumen as well as molecules embedded in the membrane. (True or False)

22. Which of the following organelles would likely have the lowest pH?
a. nucleus
b. mitochondrion
c. lysosome
d. endoplasmic reticulum
e. more than one of the above

23. How do actin and tubulin proteins relate to eukaryotic cell structure and function?
    a. They are both embedded proteins in plasma membranes.
    b. They are both components in the reactions of photosynthesis.
    c. They are both participants in the production of large amounts of ATP.
    d. They both facilitate cell motility.
    e. none of the above

24. Which of the following structures would you expect to find associated with motor proteins?
    a. nuclear pores
    b. ribosomes
    c. vesicles
    d. plasma membranes
    e. all of the above

25. Which of the following structures produces motion through the action of filaments that "slide" or "walk" parallel to each other?
    a. eukaryotic cilia
    b. eukaryotic flagella
    c. prokaryotic flagella
    d. a + b only
    e. b + c only

26. Which of the following is *not* a key difference between prokaryotic and eukaryotic flagella?
    a. Prokaryotic flagella are powered by the diffusion of hydrogen ions; eukaryotic flagella are powered by ATP.
    b. Prokaryotic flagella are composed of actin microfilaments; eukaryotic flagella are composed of microtubules.
    c. Prokaryotic flagella are used to "pull" the cell forward; eukaryotic flagella are used to "push" the cell forward.
    d. Prokaryotic flagella are capable of beating much faster than eukaryotic flagella.
    e. none of the above

Answers and Explanations

1. **d** Since elephants and oak trees are both eukaryotes, they would both require the energy production functionality of mitochondria for survival. Recall that only plant cells have cell walls and chloroplasts.

2. **b** All cells have plasma membranes, DNA, and proteins. There are single-celled representatives from both prokaryotes and eukaryotes. Of the choices given, the most reliable way to tell the difference between cell types is the presence of a nucleus, which is found only in eukaryotes.

3. **e** Electron microscopes have much greater magnifying power than light microscopes. This feature permits us to see more detail for whatever cell type we are viewing.

4. **b** All the other structures listed are eukaryotic organelles.

5. **a** The pores in the nuclear envelope permit large molecules such as proteins and RNA to enter or exit this organelle.

6. **c** Cytosol is found in prokaryotes, but not in the nucleus since this organelle would be absent.

7. **c** Vacuoles are found in both plants and fungi. They serve as a type of storage compartment in the cell. The other structures listed would not be considered true organelles.

8. **True** Proteins destined for use in the cytosol are made by free-floating ribosomes rather than by ribosomes bound to the rough ER.

9. **True** Smooth ER produces vesicles, some of which find their way to the Golgi apparatus, where additional processing of their contents occurs before those contents are directed to their final destinations.

10. **a** Lysosomes are present in animals because of their special need for internal digestion of macromolecules. Plants and fungi store some enzymes in their large vacuoles, structures that are not present in animals.

11. **e** See Figures 5.9 and 5.10 in your textbook for this information.

12. **d** The mitochondrion is the powerhouse of the cell. It is where the bulk of the reactions that manufacture ATP take place.

13. **c** The cytoskeleton supports the cell and allows for movement of the entire cell and also within the cell. All three molecules listed are part of the cytoskeleton.

14. **e** We can eliminate bacteria because they lack organelles completely. Of the organelles listed, the presence of a large vacuole only eliminates the animals. Both fungi and plants have vacuoles; therefore we cannot tell.

15. **b** The chapter refers to the mixture of lipids and proteins creating a "fluid mosaic."

16. **d** The *fluid* aspect refers to the fact that both phospholipids and proteins are somewhat mobile within the bilayer. The *mosaic* refers to the patchwork distribution of proteins across the membrane.

17. **b** The only definitive characteristic listed is the presence or absence of a nucleus (membrane-bound compartment in which the cell stores DNA).

18. **b** The division of labor within eukaryotic cells greatly increases overall cellular efficiency, just as do specialized tasks in a corporation or society.

19. **e** The ER consists of a series of interconnected tubes scattered throughout the cytosol. The connection to the nucleus allows material produced in the nucleus to directly enter the ER via the nuclear pores.

20. **d** Recall that proteins destined for the plasma membrane are produced by ribosomes found on the surface of the rough ER. Following assembly of the protein, it is then transported inside a vesicle to the Golgi apparatus, where it is processed, sorted, and sent off to its final destination (plasma membrane).

21. **True** See Figure 5.5 in your textbook for a diagram of how this occurs.

22. **c** Recall that the lysosome serves as the digestive center of the cell. It contains specialized enzymes that serve to break down various macromolecules. These enzymes operate most efficiently in an acidic environment; therefore the pH of the lysosomes is typically around 5, while the reminder of the organelles listed would have more neutral (pH 7) internal environments.

23. **d** Both actin and tubulin are proteins that make up parts of the cytoskeleton. Recall that one of the prime functions of the cytoskeleton is to facilitate the process of cell movement.

24. **c** Vesicles are transported throughout the cell by way of motor proteins that typically crawl along microtubule "tracks."

25. **d** Both eukaryotic cilia and flagella produce movement by the sliding action of parallel microtubules, which is driven by motor proteins such as dynein. Prokaryotic flagella rely on a completely different mechanism.

26. **c** Both flagella types spin or beat rapidly to produce forward propulsion of the organism.

CHAPTER 6

Cell Membranes, Transport, and Communication

Getting Started

Below are a few questions to consider prior to reading Chapter 6. These questions will help guide your exploration and assist you in identifying some of the key concepts presented in this chapter.

1. How can lichens be used to tell us about impending and potential environmental problems?
2. What properties of a phospholipid bilayer make it selectively permeable?
3. What percentage of the energy expended by a person at rest is used to fuel active transport across the plasma membrane?
4. What structures present in the cell act as "transport luggage"?
5. How many gallons of water are contained in a typical person's body?
6. What mechanisms facilitate cellular "eating" and "drinking" activities?
7. What property of steroid hormones allows them to pass easily through the plasma membrane of target cells?

A Guide to the Reading

The following concepts typically give students the most difficulty when exploring the content in Chapter 6 for the first time. For each concept, one or more references have been identified that may help you gain a better understanding of these potentially problematic areas.

Osmosis

As discussed in the chapter, the process of diffusion drives molecules to spread from areas of relatively high concentration to areas of lower concentration. In a similar fashion, when two solutions of different concentrations are separated by a selectively permeable membrane (which restricts the passage of the dissolved molecules), water molecules will exhibit directional movement. The movement of water molecules across a selectively permeable membrane is referred to as *osmosis*. The direction in which water moves across the membrane is a function of the relative concentrations of the two solutions. As an example, consider Solution A with a relatively higher concentration of dissolved molecules than Solution B. In this example, Solution A would be considered "hypertonic" relative to Solution B, since it has a higher concentration of dissolved molecules. In contrast, Solution B, with a lower concentration of dissolved molecules, would be considered "hypotonic." The key to understanding these two terms is realizing that *hypertonic* and *hypotonic* can only be used when comparing one solution to another; a single solution cannot be hypertonic or hypotonic since there would be no companion solution to compare it to. In terms of osmosis, water molecules will always move across a membrane from the hypotonic solution toward the hypertonic solution. This can actually be thought of in terms of diffusion—since the hypertonic solution, which has a higher concentration of dissolved molecules taking up space, would actually have less room for water molecules. As a result, hypertonic solutions would have relatively *fewer* water molecules than hypotonic solutions, while hypotonic solutions would have relatively *more* water molecules. If you now think of osmosis as the diffusion of water, it is easy to see why water would always move from the hypotonic solution (relatively higher concentration of water) toward the hypertonic (relatively lower concentration of water) solution. In addition, solutions that have the same concentration can be described as being *isotonic*.

For more information on this concept, be sure to focus on:

- Section 6.2, Water Requires a Cellular Balancing Act
- Figure 6.4, Water Moves Into and Out of Cells by Osmosis

Cellular Signaling

Multicellular organisms require a reliable, efficient mechanism for facilitating communication between the cells of the body. This is most typically accomplished by the release of a signaling molecule (proteins, hormones, gasses) from one cell, which is then received by a target cell, which then responds in a highly specific fashion. The types of signaling molecules and their effects on their respective target cells vary greatly. One thing to understand about signaling molecules and their target cells is the fact that in order for a target cell to receive and respond to a chemical signal, it must possess a specific receptor protein for that particular signal. These receptor proteins are typically found on the surface of the target cell and work to transmit a signal into the cell when a signaling molecule binds to it. Other classes of receptor proteins may actually be located in the cytoplasm of the cell. In this case, the signaling molecule must be permeable across the plasma membrane in order to stimulate the target cell to respond in the desired manner. Another key concept to understand about signaling molecules is that their longevity depends upon how far the molecule must travel to reach its intended target cell.

For more information on this concept, be sure to focus on:

- Section 6.6, Signaling Molecules in Cell Communication
- Figure 6.8, Receptors for Signaling Molecules

Tying It All Together

Several concepts presented in this chapter build upon concepts presented in previous chapters and are also revisited and discussed in greater detail in subsequent chapters, including:

Kingdoms of Life

- Chapter 3 – Section 3.1, The Major Groups in Context

Macromolecules

- Chapter 4 – Section 4.5, The Chemical Building Blocks of Living Systems

The Plasma Membrane

- Chapter 5 – Section 5.2, The Plasma Membrane: Separating Cells from the Environment

The Cytoskeleton

- Chapter 5 – Section 5.5, The Cytoskeleton: Providing Shape and Movement

Genes and Proteins

- Chapter 13 – Section 13.2, How Genes Control the Production of Proteins

Practice Questions

Factual Knowledge

1. Which of the following macromolecules are found in the plasma membrane?
   a. lipids only
   b. lipids and proteins
   c. lipids, proteins, and nucleic acids
   d. proteins and carbohydrates
   e. proteins only

2. Which of the following plasma membrane components is responsible for allowing the passage of charged molecules such as sodium or chloride against a concentration gradient?
   a. phospholipid bilayer
   b. channel proteins
   c. gap junctions
   d. carrier proteins
   e. receptor proteins

3. The process of dissolving sugar in a glass of water demonstrates the principle of
   a. active transport.
   b. diffusion.
   c. selective permeability.
   d. fluid mosaic.
   e. none of the above

4. Osmosis is
   a. the movement of molecules down a concentration gradient.
   b. the movement of molecules through a channel protein.
   c. the movement of large, biologically important molecules across a membrane.
   d. the movement of water across a membrane.
   e. the active transport of molecules against a concentration gradient.

5. A hypertonic solution
   a. has a lower concentration relative to another solution.
   b. has a higher concentration relative to another solution.
   c. has a concentration equal to another solution.
   d. requires the input of energy.
   e. always loses water.

6. The process of phagocytosis is considered to be a type of
   a. endocytosis.
   b. exocytosis.
   c. pinocytosis.
   d. passive transport.
   e. none of the above

7. Which of the following would *not* be considered a type of cell junction?
   a. plasmodesmata
   b. extracellular matrix
   c. gap junctions
   d. tight junctions
   e. anchoring junctions

8. Two key principles important for the evolution of large multicellular organisms are
   a. prokaryotic cell structure and cell specialization.
   b. cell specialization and communication between cells.
   c. communication between cells and simple nerve reflexes.
   d. simple nerve reflexes and cell specialization.
   e. prokaryotic cell structure and simple nerve reflexes.

9. The signaling molecules called *steroid hormones*
   a. exert their effects at a location away from the site of production.
   b. are hydrophilic and cannot penetrate the plasma membrane.
   c. bind exclusively to cell surface receptors.
   d. never enter the blood of humans.
   e. all of the above

10. Which of the following is a signaling molecule whose action promotes root formation and stem elongation in plants?
    a. nitric oxide
    b. thyroxine
    c. adrenaline
    d. platelet-derived growth factor
    e. auxin

11. Which of the following is *not* a type of signaling molecule?
    a. testosterone
    b. insulin
    c. thyhyroxine
    d. sulfur dioxide
    e. auxin

12. Match each term with the best description.
    __ osmoregulation
    __ phagocytosis
    __ plasmodesmata
    __ thyroxine
    __ hormones
    a. promotes increased metabolic rate
    b. small water transport channels
    c. long-range signaling molecules
    d. cell eating
    e. process of maintaining water balance

Conceptual Understanding

1. Consider a relatively small, complex biochemical, such as an amino acid. If this molecule passes through the plasma membrane to gain entry into the cell, it will most likely pass
   a. directly through the phospholipid bilayer.
   b. through an active carrier protein.
   c. through a channel protein.
   d. through a tight junction.
   e. though an extracellular matrix.

2. Consider the small, biologically important molecules water, oxygen, and carbon dioxide. If these molecules enter or exit the cell, they will most likely pass
   a. directly through the phospholipid bilayer.
   b. through an active carrier protein.
   c. through a channel protein.
   d. through a tight junction.
   e. through a gap junction.

3. Consider two solutions (A + B) containing a mixture of water and dissolved proteins that are separated by a membrane permeable only to water. If solution A has a relatively higher concentration of proteins,
   a. solution A would be considered hypertonic.
   b. solution B would be considered hypotonic.
   c. neither solution would be considered isotonic.
   d. all of the above
   e. none of the above

4. Consider two solutions (A + B) containing a mixture of water and dissolved proteins that are separated by a membrane permeable only to water. If solution A has a relatively higher concentration of proteins,
   a. proteins would move from solution A to solution B.
   b. water would move from solution A to solution B.
   c. proteins would move from solution B to solution A.
   d. water would move from solution B to solution A.
   e. both water and proteins would have no net movement across the membrane.

5. In our liver cells, receptors capable of recognizing LDL cholesterol particles bind to these particles and stimulate their uptake. The process by which these particles are taken into the cell is called
   a. exocystosis.
   b. phagocytosis.
   c. pinocytosis.
   d. receptor-mediated endocytosis.
   e. reverse osmosis.

6. Which of the following cell junction types would *not* allow for the passage of small signaling molecules?
   a. plasmodesmata
   b. tight junctions
   c. anchoring junctions
   d. gap junctions
   e. none of the above

7. Which of the following statements regarding signaling molecules is *false*?
   a. Signaling molecules are widely used among multicellular organisms.
   b. All signaling molecules have the same effects on their target cells.
   c. All signaling molecules require the action of specific receptor molecules present in the target cell.
   d. Signaling molecules have different longevities depending upon their activity and function.
   e. Signaling molecules have a direct effect on their target cells.

8. Refer to Table 6.1 in your textbook. Which of the following signaling molecules alters glucose uptake, and where would its receptors be located?
   a. Insulin; many different cell types that use glucose for fuel
   b. Insulin; beta cells of the pancreas
   c. PDGF; any cell that is undergoing cell division
   d. PDGF; the blood
   e. NGF; the nerves involved in simple reflexes

9. Which of the following statements regarding a typical hydrophilic signaling molecule would be *true*?
   a. Its receptor would be located in the cytosol of the target cell.
   b. Its receptor would be located on the surface of the cell.
   c. After entering the cell, it would travel directly to the nucleus.
   d. Since it can enter the cell, it directly affects specific cellular processes.
   e. It is most likely a steroid.

10. A cell is known to respond to a particular signaling molecule. Which of the following *must* be true of this cell?
    a. It is in the heart muscle.
    b. It responds to both nitric oxide and adrenaline.
    c. It is also the site of production for the signaling molecule.
    d. It contains the receptor for the signaling molecule.
    e. It is an animal cell.

11. A substance is brought to you that is a suspected steroid hormone. If this tentative identification is correct, what do you automatically know is *true* about this substance?
    a. It is likely a protein.
    b. It will likely have its cellular action very near its source of production.
    c. It will bind to and activate an intracellular receptor.
    d. It induces cell division.
    e. all of the above

12. A sudden release of adrenaline in a person undergoing a frightening experience will result in the release of stored fuels, such as the energy stored in fat. (True or False)

Related Activities

- One of the classes of signaling molecules discussed in this chapter is the steroid hormones, which are derived from cholesterol produced by the body. Using your library or the World Wide Web, perform a search to determine some of the other functions cholesterol plays in our bodies. Compose a one-page summary of your results; it should include a brief discussion of what medical conditions are associated with an overproduction of cholesterol.
- Drawing from the content presented in this chapter, search the World Wide Web to answer the question why some signaling molecules have their effects all over the body (e.g., metabolism), while others have effects only at specific locations in the body (e.g., development of sexual maturity). Using your results, compose a one-page essay describing additional examples of cell communication, besides metabolism and sexual maturation, that would fall into the categories of general or specific signaling effects.
- The discussion of acid rain presented in the Biology in the News article at the end of the chapter describes a single example of the devastating effects of acid rain precipitation and one strategy for addressing this problem. Using the World Wide Web, identify specific data sources that describe the problem of acid rain in the United States and the impact this environmental danger has had on our nation's forests. Compose a one-page summary of your results; it should include some additional strategies that are currently being implemented or should be implemented to help reduce the effects of acid rain.

Answers and Explanations

Factual Knowledge

1. **c** See Figure 6.1 in your textbook for this information. Recall that the plasma membrane consists of a phospholipid bilayer along with a series of proteins that perform specific functions for the cell, such as the channel and active/passive carrier proteins discussed in the chapter.

2. **d** Recall that active carrier proteins are used by the cell to move molecules against a concentration

gradient. These proteins require the expenditure of energy (ATP) to accomplish this feat.

3. **b** Recall that the passive (no energy required) transport of molecules down a concentration gradient illustrates the principle of diffusion.

4. **d** Osmosis is the movement of water across a membrane. The direction of movement depends on the relative concentrations of the solutions on either side of the membrane.

5. **b** Recall that the terms *hypotonic*, *hypertonic*, and *isotonic* can only be used when comparing two or more solutions. A hypertonic solution always has a relatively higher concentration.

6. **a** Phagocytosis, also called "cell eating," involves the cell engulfing large particles. This process is a type of endocytosis.

7. **b** While the extracellular matrix plays a role in binding cells together, it does not contribute to the overall communication between cells, a primary function of cell junctions.

8. **b** Multicellular organisms cannot be prokaryotic, and not all large multicellular organisms have nervous systems. Cell communication and specialization are the two main organizing principles discussed in this chapter.

9. **a** Steroid hormones are derived from cholesterol and can therefore pass through membranes to also bind with receptors in the cytosol. They travel all over the body, mainly by way of the blood, and have target cells that can be located a considerable distance from the source of hormone production.

10. **e** Auxin, an acetic acid derivative found in most plant cells, performs this function.

11. **d** Sulfur dioxide is an environmental contaminant that contributes to the development of acid rain.

12. **e** osmoregulation
d phagocytosis
b plasmodesmata
a thyroxine
c hormones

Conceptual Understanding

1. **c** Recall that channel proteins located in the plasma membrane serve to facilitate the transport of smaller complex biochemicals, such as amino acids, sugars, and small proteins.

2. **a** Recall that small molecules such as water, oxygen, and carbon dioxide can pass directly through the plasma membrane without the expenditure of energy. This makes the transport of these biologically important molecules easy and efficient for cells.

3. **d** The terms *hypertonic*, *hypotonic*, and *isotonic* are used to describe the relative concentrations of two solutions. In this case, choices a, b, and c are all accurate descriptions.

4. **b** The terms *hypertonic*, *hypotonic*, and *isotonic* are used to describe the relative concentrations of two solutions. In this case, the process of osmosis would draw water from solution A into Solution B across the membrane.

5. **d** The process of endocytosis is used by the cell to import external compounds. In receptor-mediated endocytosis, specialized receptor proteins are responsible for identifying and determining which substances, such as LDL cholesterol, enter the cell.

6. **b** Recall that tight junctions are responsible for holding cells together using strands of rigid proteins. Tight junctions form leak-proof barriers that prevent the passage of molecules. Refer to figure 6.7 for additional information.

7. **b** The activity and effect of a signaling molecule on its target cell are highly variable and depend on the nature of both the signal and the receptor involved in the response.

8. **a** Although insulin is produced by the beta cells of the pancreas, it promotes glucose uptake in many different body cell types.

9. **b** Hydrophilic molecules cannot cross membranes, so the way this signaling molecule would have its effect is by interacting with a cell surface receptor.

10. **d** Only target cells with appropriate receptors can respond to signaling molecules.

11. **c** Since steroid hormones, such as progesterone, are derived from cholesterol and are hydrophobic, they pass directly through the plasma membrane and bind to and activate intracellular receptors. Once activated, these steroid-protein complexes stimulate the production of specific proteins that are necessary for the target cell's response.

12. **True** This effect of adrenaline makes possible the conversion of fats into glucose for energy.

CHAPTER 7 Energy and Enzymes

Getting Started

Below are a few questions to consider prior to reading Chapter 7. These questions will help guide your exploration and assist you in identifying some of the key concepts presented in this chapter.

1. How do gasoline-powered hedge trimmers demonstrate the principle of the first law of thermodynamics?
2. How does an aging tool shed demonstrate the principle of the second law of thermodynamics?
3. What is the primary energy source for all living organisms?
4. What is a bomb calorimeter, and what does it measure?
5. How is the activation energy for a chemical reaction similar to the sides of a basket holding a litter of puppies?
6. What are enzymes, and what function do they play in biochemical reactions?
7. How might metabolism shorten the lifespan of living organisms?
8. How many minutes of energetic dancing would be needed to burn the number of calories consumed from a large hamburger and fries?

A Guide to the Reading

The following concepts typically give students the most difficulty when exploring the content in Chapter 7 for the first time. For each concept, one or more references have been identified that may help you gain a better understanding of these potentially problematic areas.

Laws of Thermodynamics

Thermodynamics is a dedicated field of physics that studies the dynamics of heat (from the Greek term *thermo*). Heat is a form of energy that is involved in almost all chemical reactions. There are certain universal laws that govern how energy behaves in all chemical reactions; these are the laws of thermodynamics. As discussed in the chapter, the *first law of thermodynamics* dictates that energy cannot be created (or destroyed), but can only be transformed from one form to another. An excellent example of this is the gasoline used to power automobiles. Gasoline contains energy in a liquid form that is stable at room temperature. When gas is burned inside an engine, the energy is released. This energy takes on different forms—force (used to power the movement of the pistons, propelling the car forward) and heat. In this example, the amount of energy contained in the gasoline exactly equals the amount of energy utilized as force and lost as heat. In this and all cases, energy is not destroyed or created—it is merely transformed from one form (liquid gasoline) to another (force and heat). The *second law of thermodynamics* dictates that all systems (i.e., a cell, a bedroom, a yard, the biosphere) always tend to become more disordered. An excellent example of this would be your dormitory room. Over time, the room will always become more messy (disordered). The only way to reverse this trend is through the expenditure of energy (cleaning your room). Without this input of energy to restore order to the system, it will remain in a perpetual state of disorder. The use of energy in a cell is governed by these two laws of thermodynamics without exception. The multitude of chemical reactions that provide the cell's metabolism must always follow these rules.

For more information on this concept, be sure to focus on:

- In Section 7.1, The laws of thermodynamics apply to living systems
- Figure 7.1, The Second Law of Thermodynamics

Oxidation and Reduction Reactions

The process of electron transfer plays a crucial role in the multitude of chemical reactions in which cells utilize, capture, and store energy. The transfer of electrons between molecules will result in one molecule gaining electrons (reduction) and the other molecule losing electrons (oxidation). It is important to note that oxidation and reduction often occur as a paired process. Because of this close association, the textbook indicates these reactions are often called *oxidation-reduction*, or *redox*, reactions. The loss of electrons is called *oxidation* because when carbon atoms lose electrons through this process, they are often bound to oxygen atoms as a result. Therefore, while the carbon atom has lost electrons, it has gained oxygen atoms—hence the term *oxi-dation.* Conversely, compounds that are capable of accepting electrons will also tend to lose oxygen atoms, leaving them in a reduced state; hence the term *reduction.* Oxidation-reduction reactions play a huge role in the capture of energy, in the form of sugar during photosynthesis and ATP during respiration.

For more information on this concept, be sure to focus on:

- In Section 7.2, Capturing energy from foods requires the transfer of electrons

Enzyme Catalysis

As described in the chapter, all chemical reactions require a certain amount of activation energy in order to proceed. The activation energy required prevents the reaction from proceeding without regulation. For many reactions, however, the activation energy can be prohibitive. Cells have found a way to overcome this limitation by employing a special class of proteins, called *enzymes,* which work to catalyze, or speed up, a chemical reaction. Catalysis is accomplished by lowering the amount of activation energy required to drive the reaction forward. Enzymes can accomplish this in several ways, including by binding to the substrates of the reaction, bringing them in close proximity to one other, or placing stress on chemical bonds. The key concept to remember when discussing enzymes is that, while they play a key role in driving chemical reactions, they are not consumed during the reaction. This means a single enzyme molecule can catalyze the same reaction multiple times, remaining unchanged throughout the process. Another concept to keep in mind is that enzymes are highly specific—they typically drive one, and only one, chemical reaction.

For more information on this concept, be sure to focus on:

- In Section 7.3, Enzymes speed up chemical reactions
- In Section 7.3, The shape of an enzyme directly determines its activity
- Figure 7.6, Enzymes as Molecular Matchmakers

Tying It All Together

Several concepts presented in this chapter build upon concepts presented in previous chapters and are also revisited and discussed in greater detail in subsequent chapters, including:

Chemical Structures Found in Cells

- Chapter 4 - Section 4.5, The Chemical Building Blocks of Living Systems

ATP: Energy for the Cell

- Chapter 4 - in Section 4.5, Nucleotides store information and energy

Chloroplast and Energy Capture

- Chapter 5 - in Section 5.4, Chloroplasts capture energy from sunlight
- Chapter 8 - Section 8.2, Photosynthesis: Capturing Energy from Sunlight

Mitochondria and Energy Production

- Chapter 5 - in Section 5.4, Mitochondria are the power stations of the eukaryotic cell
- Chapter 8 - Section 8.3, Catabolism: Breaking Down Molecules for Energy

DNA Synthesis

- Chapter 12 - Section 12.2, The Three-Dimensional Structure of DNA

Carbon Fixation and Cycling

- Chapter 25, Global Change

Practice Questions

Factual Knowledge

1. Consider the gas-powered generator used to run the electric hedge clipper discussed in the chapter. The use of one form of energy (gasoline) to drive a process dependent upon a different form of energy (electricity) illustrates
   a. the first law of thermodynamics.
   b. the second law of thermodynamics.
   c. both laws of thermodynamics.
   d. neither law of thermodynamics.
   e. metabolism.

2. Metabolic reactions that break down complex molecules into smaller compounds, thereby releasing usable energy for the cell, are best described as
   a. biosynthetic.
   b. catabolic.
   c. catalytic.

d. photosynthetic.
e. enzymatic.

3. The ultimate source of energy for living systems is
   a. glucose.
   b. oxygen.
   c. sunlight.
   d. carbon dioxide.
   e. none of the above

4. Which of the following statements about cellular energy needs is *false*?
   a. Cell disorder will always tend to increase without a continuous input of energy.
   b. Since the laws of thermodynamics dictate that energy cannot be created or destroyed, this forces cells to acquire energy from external forces.
   c. Many cellular reactions have an associated activation energy.
   d. The most usable energy for cells comes directly from the rapid combustion of glucose.
   e. all of the above

5. Since animals cannot obtain carbon directly from the atmosphere, or acquire energy directly from the sun, they ultimately rely on the metabolism of plants to meet these critical needs. (True or False)

6. Biological oxidation is most useful to cells if it occurs in ways that make large amounts of energy available all at once. (True or False)

7. The production and breakdown of _______ is often coupled with the metabolic reactions of biosynthesis and catabolism.
   a. aspirin
   b. DNA
   c. ATP
   d. methane
   e. CO_2

8. Which of the following statements about enzymes is *true*?
   a. Enzymes do not alter the overall change in free energy for a reaction.
   b. Enzymes are proteins whose three-dimensional shape is key to their function.
   c. Enzymes speed up reactions by lowering activation energy.
   d. Enzymes are highly specific for the reactions they catalyze.
   e. all of the above

9. Carbonic anhydrase is an enzyme that speeds the capture of carbon dioxide in photosynthesis, thus facilitating carbon transfer from plants to animals. (True or False)

10. One strategy that makes metabolic pathways more efficient is
    a. that enzymes for a given pathway are located in the same place within the cell.
    b. that all of the reactions use the same substrate.
    c. that all of the reactions use the same enzyme.
    d. that enzymes for a pathway are separated into different organelles.
    e. none of the above

11. Which of the following is *not* involved in the biosynthesis of DNA?
    a. energy from ATP
    b. mononucleotides
    c. carbonic anhydrase
    d. enzymes
    e. all of the above

12. The enzyme COX-1 is important in human health because it
    a. helps transport carbon dioxide in the blood.
    b. is critical for the biosynthesis of DNA.
    c. is a chemical derivative of aspirin.
    d. catalyzes the production of hormones that maintain the stomach lining.
    e. catalyzes the production of hormones that promote inflammation and fever.

13. Match each term with the best description.
    __ active site
    __ catalyst
    __ metabolism
    __ reduction
    __ substrate
    a. capture and use of energy in organisms
    b. converted into product by an enzyme
    c. critical three-dimensional space on an enzyme
    d. speeds up the rate of a reaction
    e. result of molecule gaining electrons

Conceptual Understanding

1. The medicinal effects of aspirin have their biochemical basis in the inhibition of an enzyme associated with inflammation and fever. (True or False)

2. Which of the following would be considered as metabolism?
   a. biosynthetic pathways that build DNA
   b. catabolic pathways that break down complex carbohydrates
   c. the capture of light energy for use in making glucose
   d. the manufacture of ATP from the breakdown of glucose
   e. all of the above

3. Your dorm room or apartment will stay organized only if you expend sufficient energy to keep it straightened up. Likewise, cells can remain in a highly ordered state only if they have a supply of energy to use for cellular work. These two examples clearly illustrate
   a. the first law of thermodynamics.
   b. the second law of thermodynamics.
   c. reduction and oxidation.
   d. enzyme specificity.
   e. all of the above

4. Refer to Figure 7.2 in your textbook. In this diagram, organisms such as the wolf are
   a. direct consumers of CO_2.
   b. indirect consumers of CO_2.
   c. direct producers of CO_2.
   d. indirect producers of CO_2.
   e. indirect consumers of CO_2 and both direct and indirect producers of CO_2.

5. A common way in which cells capture the energy released during the breakdown of large molecules is to add electrons to smaller, specialized molecules that can accept them. This process of electron acceptance is otherwise known as
   a. biosynthesis.
   b. metabolism.
   c. reduction.
   d. catalysis.
   e. a pathway.

6. In Figure 7.5 of your textbook, an analogy is drawn between frisky puppies in a basket and the activation energy of an enzyme-regulated reaction. Extending this analogy, what would be the effect of lowering the sides of the basket?
   a. The puppies would start rolling down the hill more easily.
   b. The puppies would roll down the hill more slowly.
   c. Some of the puppies would now go back up the hill.
   d. The puppies would have a much harder time getting out of the basket.
   e. Fewer puppies would end up at the bottom of the hill.

7. Before they can react, many molecules must first be destabilized. This state is typically achieved through
   a. changing the chemical composition of these molecules.
   b. oxidizing the molecules by adding electrons.
   c. changing the reaction from a biosynthetic to a catabolic pathway.
   d. the input of a small amount of activation energy.
   e. none of the above

8. In Figure 7.6a, the object labeled *enzyme* is
   a. a three-dimensional protein.
   b. one component of a larger metabolic pathway.
   c. involved in lowering the activation energy of the reactants to which it binds.
   d. unchanged by its participation in the reaction.
   e. all of the above

9. In the enzyme-catalyzed reaction where A → E1 → B → E2 → C → E3 → D, what will be the effect of inactivating the enzyme labeled E2?
   a. A, B, C, and D will still all be produced.
   b. A, B, and C will still be produced, but not D.
   c. A and B will still be produced, but not C or D.
   d. A will still be produced, but not B, C, or D.
   e. None of the labeled substances will be produced.

10. Enzyme-driven metabolic pathways can be made more efficient by
    a. concentrating enzymes within specific cellular compartments.
    b. grouping enzymes into free-floating, multi-enzyme complexes.
    c. fixing enzymes into membranes so that they are adjacent to each other.
    d. having the product of one reaction become the reactant of the next.
    e. all of the above

11. The process of biosynthesis diagrammed in Figure 7.8 of your textbook
    a. is used to make a number of different kinds of macromolecules.
    b. can occur in the absence of ATP.
    c. does not require enzymes.
    d. can take place in a single chemical step.
    e. all of the above

12. Metabolic rate has been shown to be related to the lifespan of organisms. Which of the following statements regarding this observation is *false*?
    a. Lower metabolic rates are associated with a longer life span.
    b. Higher metabolic rates are associated with a higher production of toxic by-products.
    c. Metabolic rate can be controlled by restricting the organism's diet.
    d. In general, smaller organisms have slower metabolisms than larger organisms.
    e. Genetics, along with metabolism, plays a role in determining lifespan.

Related Activities

- Refer to the Biology in the News article "More Heart Risks Found for COX-2 Inhibitors" at the end of the chapter. Search the World Wide Web for additional information on COX-2 inhibitors and compose a one-page summary of the reported benefits of these "superaspirins" and the current research that has shown the dangers associated with their prolonged use.
- The Biology Matters box titled "Energy Is a Weighty Subject" discusses how metabolism relates to one's waistline. In this section is the statement "If you take in more calories than your BMR consumes, you will gain weight, unless you increase your activity level to use up the extra calories." Compare this concept with the information presented in Section 7.5, "Metabolic Reactions and Life Span." Compose a one-page summary weighing the benefits and drawbacks of a restricted diet for weight loss when compared to a rigorous exercise program.
- Whenever you cut a potato, apple, or other fresh fruit or vegetable, before long the exposed tissue will turn brown. This color change is mediated by an enzyme in the plant tissue called *catecholase*, which converts a colorless substrate known as catechol into a colored product called *benzoquinone* using an oxidation reaction. One may prevent cut fruit from browning by covering it with lemon juice or submerging it in water. Suggest a biological explanation that involves enzyme function to explain why this preservation technique works.

Answers and Explanations

Factual Knowledge

1. **a** The gas in the generator can be converted to electricity and used to run the hedge clippers, with some resulting heat given off as well. This example demonstrates energy conversions governed by the first law of thermodynamics.
2. **b** This is the definition of catabolic reactions.
3. **c** Although glucose is used by cells to produce ATP, it is ultimately sunlight that provides the energy needed to make glucose through the process of photosynthesis.
4. **d** It is the slow, stepwise combustion of glucose that makes the harvest of energy so efficient. This process provides the cell with most of its usable energy.
5. **True** Plants capture both sunlight energy and carbon in the form of CO_2 during photosynthesis. These are then passed along to animals in the form of glucose and other macromolecules.
6. **b** The cell works best with small increments of energy from oxidation.
7. **c** ATP is the currency of metabolism. It is frequently made or used in conjunction with other metabolic reactions.
8. **e** All of these statements about enzymes are true.
9. **False** The action of carbonic anhydrase has nothing to do with photosynthesis. It is an enzyme that catalyzes the conversion of water and carbon dioxide into bicarbonate ions in your blood.
10. **a** Concentrating both substrates and enzymes in the same cellular location is one way to make metabolic pathways more efficient. Other ways include attaching enzymes to membranes or combining them into free-floating multi-enzyme complexes.
11. **c** Carbonic anhydrase is involved in the process that transports carbon dioxide in the blood.
12. **d** The first two answers have nothing to do with COX enzymes. Aspirin is not a derivative of COX-1, but it does inhibit the normal function of COX enzymes. COX-2 is involved with inflammation and fever responses.
13. **c** active site
    d catalyst
    a metabolism
    e reduction
    b substrate

Conceptual Understanding

1. **True** The enzyme is COX-2, which promotes production of a hormone that causes inflammation and fever. Inhibiting COX-2 with aspirin relieves these conditions.
2. **e** All of these are ways in which energy is either captured or used by cells and thus qualify as part of metabolism.
3. **b** The second law of thermodynamics says that the disorder of a system will increase without an input of energy, whether that system be a cell or your room.
4. **e** The wolf indirectly consumes CO_2 by eating animals that have eaten plants, which are the direct consumers. The wolf is both a direct producer of CO_2 through respiration, and an indirect producer through the decay of its feces or of its body when it dies.

5. **c** Reduction is the transfer of energy through the acceptance of electrons by a molecule. It is the opposite of oxidation, in which electrons are released.

6. **a** This is the equivalent of lowering the activation energy for the reaction, which means the reaction will occur more easily (that is, more puppies can roll down the hill for a given amount of energy). This is what enzymes do to make reactions proceed at a faster rate.

7. **d** This small amount of added energy is usually sufficient to overcome the activation energy barrier and destabilize the molecule.

8. **e** All of these statements about enzymes are true.

9. **c** Since E2 catalyzes the conversion of B into C, no C will be made. This means that E3 cannot convert C into D, so no D will be made either. However, since E1 is still functional, A will be converted into B.

10. **e** All of these are mechanisms used by cells to increase the efficiency of enzymatic pathways in metabolism.

11. **a** The same general process of using ATP to energize monomers is the basis for the biosynthesis of a wide variety of polymers.

12. **d** Recall that smaller animals generally have faster metabolic rates and, therefore, shorter life spans than larger animals.

CHAPTER 8 Photosynthesis and Respiration

Getting Started

Below are a few questions to consider prior to reading Chapter 8. These questions will help guide your exploration and assist you in identifying some of the key concepts presented in this chapter.

1. In terms of energy requirements, what is the consequence of humans' brain size to body weight ratio?
2. What is phosphorylation, and how does it affect proteins?
3. Why are the leaves of most plants colored green?
4. How are proton gradients used in cells to generate chemical energy?
5. Under what conditions are the products ethanol and lactic acid produced by organisms?
6. What occurs during the process of developing bread dough?

A Guide to the Reading

The following concepts typically give students the most difficulty when exploring the content in Chapter 8 for the first time. For each concept, one or more references have been identified that may help you gain a better understanding of these potentially problematic areas.

Phosphorylation

The energy carrier molecule ATP is often used to transfer energy to organic compounds within the cell. When one of the phosphate atoms from the ATP molecule is transferred to an organic compound, this compound is said to be "phosphorylated." Phosphorylation of molecules can result in changes in their shape or activation/deactivation of enzymatic activity. This process plays a key role in regulating the activities of enzymes within the cell. It is important to note that the cell utilizes a process called *oxidative phosphorylation* to regenerate ATP during aerobic respiration. In this process, the energy stored in the proton gradient established across the inner mitochondrial membrane is used to drive the activity of the enzyme ATP synthase. ATP synthase is responsible for phosphorylating ADP molecules to generate ATP.

For more information on this concept, be sure to focus on:

- Section 8.1, Energy Carriers: Powering All Activities of the Cell
- In Section 8.3, Oxidative phosphorylation uses oxygen and NADH to produce ATP in quantity
- Figure 8.11, Oxidative Phosphorylation

Light vs. Dark Reactions of Photosynthesis

Photosynthesis, which occurs in the chloroplast, can be broken down into two interrelated processes—light reactions and dark reactions. The light reactions of photosynthesis work to capture the energy of sunlight inside the chloroplast. As discussed in the chapter, the pigment chlorophyll is green in color—this is because it reflects green light, causing our eyes to see green when we look at a leaf. Consequently, the pigment is efficient at absorbing both red-orange and blue-violet light. When light is absorbed, electrons present in the chlorophyll molecule become "energized" and are then passed on to an electron transport chain (ETC) that is arranged in one of two types of photosystems (I and II). Photosystem I accepts high energy electrons and produces NADPH, a compound used in the cell to reduce compounds. Photosystem II accepts electrons and produces the high energy molecule ATP. In either case, the transport of electrons in the ETC helps to generate a proton gradient across the thylakoid membrane in the chloroplast, which can then be used to generate ATP in the cell. The ATP and NADPH produced by the light reactions of photosynthesis are utilized by the cell to drive the dark reactions of photosynthesis. In the dark reactions, carbon dioxide from the air is combined with water to produce glucose molecules and oxygen. This process requires energy obtained from both ATP and hydrogen ions donated by NADPH. As indicated by its name, the dark reactions of photosynthesis do not require light in order to proceed.

For more information on this concept, be sure to focus on:

- In Section 8.2, The light reactions capture energy from sunlight
- In Section 8.2, The dark reactions manufacture sugars
- Figure 8.4, The Arrangement of Photosystems in the Thylakoid Membrane
- Figure 8.5, Production of Energy Carriers by the Light Reactions
- Figure 8.6, The Dark Reactions Fix Carbon

The Citric Acid Cycle

During aerobic respiration, the process of glycolysis is followed by a stepwise series of reactions that serve to donate high energy electrons to the compound NAD^+, generating NADH. As discussed in the chapter, the process of glycolysis results in the generation of two molecules of pyruvate for every glucose molecule used. The three-carbon pyruvate releases a single CO_2 molecule and joins to a compound called coenzyme A. The resulting molecule is called acetyl CoA. Acetyl CoA then enters the citric acid cycle, which is a series of eight oxidation reactions that occur within the mitochondrial matrix. As the citric acid cycle proceeds, carbon atoms are released as CO_2 (which are then exhaled) and high energy electrons are stored in NADH. The NADH molecules produced from the citric acid cycle are then used to drive the final stage of respiration—oxidative phosphorylation—which produces ATP. It is important to note that very little ATP is produced during the citric acid cycle. The primary product is in fact the high energy NADH molecules.

For more information on this concept, be sure to focus on:

- In Section 8.3, The Krebs cycle produces NADH and carbon dioxide
- Figure 8.10, The Citric Acid Cycle

Tying It All Together

Several concepts presented in this chapter build upon concepts presented in previous chapters and may also be revisited and discussed in greater detail in subsequent chapters, including:

Mitochondrial and Chloroplast Structure

- Chapter 5 – Section 5.4, The Specialized Internal Compartments of Eukaryotic Cells

Diffusion

- Chapter 6 – Section 6.1, The Plasma Membrane Is Both Gate and Gatekeeper

Oxidation-Reduction Reactions

- Chapter 7 – in Section 7.2, Capturing energy from foods requires the transfer of electrons

Metabolism

- Chapter 7 – Section 7.2, Using Energy from the Controlled Combustion of Food

Practice Questions

Factual Knowledge

1. Which of the following statements about energy metabolism is *false*?
   a. The energy that powers living systems ultimately comes from the sun.
   b. All animals in some way rely on plants for their energy.
   c. Plants provide the water and CO_2 that animals need to carry out respiration.
   d. Most eukaryotic organisms carry out respiration in the presence of O_2.
   e. The main objective of respiration is to break down glucose and make ATP.

2. Which of the following is the *reduced* form of a temporary electron carrier molecule?
   a. NADH
   b. ATP
   c. $NADP^+$
   d. CO_2
   e. all of the above

3. The majority of energy carrier molecules are oxidized or reduced in the
   a. nucleus.
   b. Golgi apparatus.
   c. cytosol.
   d. mitochondria and chloroplasts.
   e. all of the above

4. The stroma is located in the mitochondrion and is the place where the light reactions of photosynthesis take place. (True or False)

5. Which of the following are logically associated with chloroplasts?
   a. plant cells
   b. dark reactions
   c. chlorophyll
   d. thylakoid membranes
   e. all of the above

6. Antenna complexes, electron transport chains, and carbon fixation are all found in
   a. animal cells.
   b. bacterial cells.
   c. plant cells.
   d. mitochondria.
   e. all of the above

7. The manufacture of ATP in both photosynthesis and respiration is made possible by
   a. the existence of a proton gradient across specific membranes.
   b. the action of ATP synthase.
   c. energy from the movement of electrons.
   d. the action of electron-accepting proteins.
   e. all of the above

8. Rubisco, glyceraldehyde 3-phosphate, and NADPH all play a role in
   a. the light reactions of photosynthesis.
   b. the dark reactions of photosynthesis.

c. the breakdown of glucose into CO_2.
d. cellular respiration when O_2 is present.
e. alcohol fermentation.

9. Six molecules of CO_2 must be captured during the dark reactions of photosynthesis for every molecule of glucose that is manufactured. (True or False)

10. Which of the following represents the correct ordering of events that occur during the catabolism of glucose in the *absence* of O_2?
    a. glycolysis; citric acid cycle; oxidative phosphorylation
    b. glycolysis; oxidative phosphorylation; citric acid cycle
    c. oxidative phosphorylation; citric acid cycle; glycolysis
    d. glycolysis; fermentation
    e. citric acid cycle; fermentation

11. Glycolysis takes place in the ________ and produces ________, which in the presence of oxygen then enters the ________.
    a. cytosol; glucose; mitochondrion to complete fermentation
    b. cytosol; pyruvate; mitochondrion to complete fermentation
    c. cytosol; pyruvate; mitochondrion to complete cellular respiration
    d. mitochondrion; pyruvate; chloroplast to complete photosynthesis
    e. chloroplast; glucose; cytosol to complete cellular respiration

12. During cellular respiration, ________ is the step that produces the greatest number of ATP molecules.
    a. fermentation
    b. glycolysis
    c. citric acid cycle
    d. oxidative phosphorylation
    e. carbon fixation

13. Both NAD^+ and $NADP^+$
    a. are reduced.
    b. have a full complement of electrons.
    c. are oxidized.
    d. are used during carbon fixation in photosynthesis.
    e. none of the above

14. Match each term with the best description.
    __ aerobic
    __ catabolism
    __ energy carrier
    __ glycolysis
    __ thylakoids
    a. in the presence of oxygen
    b. takes place in the cytosol
    c. cellular respiration is an example
    d. embedded within the stroma
    e. NADH

Conceptual Understanding

1. Refer to Figure 8.1 in your textbook. What is the overall purpose of the two sets of metabolic processes shown?
   a. to make glucose out of sunlight energy
   b. to use sunlight energy to manufacture ATP
   c. to fix carbon into biological molecules
   d. to make O_2 so that organisms can respire
   e. to recycle CO_2 between plants and animals

2. Which of the following processes ultimately provides the O_2 used during the process of oxidative phosphorylation?
   a. fermentation
   b. light reactions of photosynthesis
   c. dark reactions of photosynthesis
   d. citric acid cycle
   e. carbon fixation

3. Imagine that you have been shrunk to microscopic size and are sitting on top of a thylakoid disk. Which of the following are contained in the structure upon which you sit?
   a. photosystems
   b. antenna complexes
   c. electron transport chains
   d. phospholipids
   e. all of the above

4. Which of the following serves as both reactant in photosynthesis and product in cellular respiration?
   a. O_2
   b. CO_2
   c. H_2O
   d. a and b
   e. b and c

5. The protons that make up the proton gradient used during the light reactions of photosynthesis come from
   a. glucose.
   b. ATP.
   c. H_2O.
   d. NADPH.
   e. rubisco.

6. The electrons that are released by the splitting of water molecules during photosynthesis ultimately end up in
   a. glucose.
   b. ATP.
   c. H_2O.
   d. NADPH.
   e. rubisco.

7. What process in cellular respiration is essentially the reverse of carbon fixation in photosynthesis?
   a. glycolysis
   b. citric acid cycle
   c. oxidative phosphorylation
   d. alcohol fermentation
   e. lactic acid fermentation

8. In order to carry out glycolysis, a cell must have glucose and mitochondria. (True or False)

9. Acetyl CoA, CO_2, oxaloacetate, and NADH are all
   a. components or products of the citric acid cycle.
   b. part of the dark reactions of photosynthesis.
   c. part of the reactions of lactic acid fermentation.
   d. elements of oxidative phosphorylation.
   e. all of the above

10. The muscle cells of white lab rats have 50 percent more mitochondria than do those of burrowing blind naked mole-rats. (True or False)

11. If you consider the combined processes of photosynthesis and cellular respiration, the electrons found in H_2O at the beginning of the light reactions end up attaching to _______ during respiration.
    a. O_2 to make new H_2O
    b. NADPH to make new glucose
    c. pyruvate to make ethanol
    d. electron transport carriers to make O_2
    e. none of the above

12. Consider a hypothetical eukaryotic cell that has only a single glucose molecule available for respiration. If this cell can only perform fermentation instead of aerobic respiration, it
    a. will produce fewer CO_2 molecules as a result.
    b. will generate fewer ATP molecules as a result.
    c. probably lacks O_2.
    d. will have mitochondria present.
    e. all of the above

13. Oxidative phosphorylation is to respiration as _______ is to photosynthesis.
    a. carbon fixation
    b. the splitting of water
    c. the electron transport chain
    d. light capture by chlorophyll
    e. the reduction of NADPH

Related Activities

- The information regarding the "brain drain" presented at the end of the chapter suggests that the dietary intake of a mother plays a major role in brain development for an unborn child. Search the World Wide Web or your library for additional information about this nutritional link between mother and child during pregnancy. Compose a one-page summary of what you find. In particular, take note of the potentially debilitating effects on the fetus of nutritional deficits in a mother's diet.
- Each of your body cells uses up its store of ATP every minute or two. It should therefore be apparent to you just how critical the role of oxidative phosphorylation is to your body, since that is where the vast majority of ATP comes from. With this in mind, search the World Wide Web for information on the fast-acting poison called *cyanide*. Compose a one-page essay discussing the relationship between mitochondria, oxidative phosphorylation, and the action of cyanide.
- Consider the story "Biodiesel Boom Well-Timed" in the Biology in the News section at the end of the chapter. Compose a brief editorial to your local newspaper that (1) outlines the biology behind this alternative fuel source, and (2) persuasively argues for widespread community adoption of biodiesel as an alternative to fossil fuels.

Answers and Explanations

Factual Knowledge

1. **c** Respiration does not use water or CO_2 as reactants; these are actually two of the products of respiration. Plants provide the glucose and O_2 that make respiration possible.
2. **a** ATP and CO_2 are not electron carriers, and $NADP^+$ is an oxidized form of a carrier.
3. **d** These two organelles are the prime players in energy metabolism. A few electron carriers are also reduced in the cytosol during glycolysis.
4. **False** This statement is wrong on both counts. The stroma is part of the chloroplast, and it is where the dark reactions of photosynthesis take place.
5. **e** Chloroplasts are found in plant cells and contain both thylakoid membranes and chlorophyll. The dark reactions take place within the stroma of the chloroplast.
6. **c** These are all participants or processes associated with photosynthesis, which occurs in plant cells.
7. **e** As electrons move through the electron acceptor molecules present on the inner membranes of both chloroplasts and mitochondria, the energy associated with the electrons is used to generate a proton gradient. Protons are then passed through an ATP synthase to provide the energy needed to manufacture ATP.

8. **b** All three are players in the dark reactions of photosynthesis, where carbon fixation occurs. Rubisco is an enzyme that catalyzes the capture of CO_2, glyceraldehyde 3-phosphate is one of the intermediates in the chemical pathway that leads to glucose, and NADPH is an electron carrier made during the light reactions that is used to power the dark reactions.

9. **True** There are six carbons in a glucose molecule, so six molecules of CO_2 must be captured for each glucose made.

10. **d** In the absence of O_2, all that can occur is glycolysis and fermentation. O_2 must be present for the citric acid cycle and oxidative phosphorylation to take place.

11. **c** See Figures 8.8 and 8.10 in your textbook for a summary of these possibilities.

12. **d** Over 80 percent of the ATP produced during cellular respiration is the result of oxidative phosphorylation.

13. **c** The positive sign indicates a charge that results from the loss of electrons, which is what makes these molecules oxidized.

14. **a** aerobic
c catabolism
e energy carrier
b glycolysis
d thylakoids

Conceptual Understanding

1. **b** All of these things happen as a result of some aspect of energy metabolism. However, the overall purpose of the combined processes of photosynthesis and respiration is to take an abundant, but unusable, form of energy (sunlight) and convert it to a form of energy that is usable by the cell (ATP).

2. **b** O_2 is made when water molecules are split to generate electrons during the light reactions of photosynthesis.

3. **e** The thylakoid disk is part of the inner membrane of the chloroplast and is involved with the light reactions of photosynthesis. Therefore, all of these structures would be a part of the disk.

4. **e** Both H_2O and CO_2 are used during photosynthesis to help make glucose, and are produced during the breakdown of glucose that occurs during respiration.

5. **c** The splitting of water molecules in the light reactions gives rise to the protons needed for this gradient.

6. **d** The reason for splitting water molecules is to provide electrons to the photosystems used during the light reactions. These photosystems need new electrons because some of them are used to reduce $NADP^+$ into NADPH.

7. **b** Carbon fixation converts CO_2 into glucose, and the citric acid cycle essentially reverses this process.

8. **False** Glycolysis occurs in the cytosol of the cell. See Figure 8.8 in your textbook for this information.

9. **a** See Figure 8.10 in your textbook for this information.

10. **False** Just the reverse is true. Naked mole-rats live in oxygen-poor underground tunnels. They spend much of their time digging through hard soils, and therefore have a great demand for ATP. To generate the necessary ATP in a low O_2 environment, the muscle cells of the mole-rat have adapted by accumulating more mitochondria as compared to lab rats.

11. **a** Interestingly, the electrons both begin their journey through metabolism in water and end that journey in the water that is made when oxygen serves as the final electron acceptor in oxidative phosphorylation, before combining with protons to form new "metabolic" water.

12. **e** The lack of O_2 is what directs the products of glycolysis into fermentation instead of into the mitochondrion for use in aerobic respiration. This will result in reduced CO_2 and ATP production.

13. **c** Both processes use electron acceptor molecules to carry out a series of oxidation-reduction reactions that generate energy for use in the production of ATP.

CHAPTER 9 Cell Division

Getting Started

Below are a few questions to consider prior to reading Chapter 9. These questions will help guide your exploration and assist you in identifying some of the key concepts presented in this chapter.

1. How does the microorganism *Borrelia*, the causative agent of Lyme disease, cause the body's immune system to undergo an identity crisis?
2. What are the differences between the G_1 and G_2 phases of the cell cycle?
3. What role do the cytoskeletal structures called *microtubules* play in cell division?
4. What important role do vesicles play in the process of plant cell division?
5. What is the primary difference between haploid and diploid cells?
6. How many copies of each chromosome are present in cells that have undergone meiosis I? How about meiosis II?

A Guide to the Reading

The following concepts typically give students the most difficulty when exploring the content in Chapter 9 for the first time. For each concept, one or more references have been identified that may help you gain a better understanding of these potentially problematic areas.

Homologous Chromosomes

As described in the chapter, homologous chromosomes are pairs of chromosomes containing the same set of genes. One homolog of the pair is inherited from the organism's mother, while the other homolog is inherited from the father. Humans possess 23 pairs of homologous chromosomes for a grand total of 46 individual chromosomes. The process of cell division must utilize a mechanism that ensures that each daughter cell obtains a complete set of 23 pairs of homologous chromosomes. To accomplish this, each individual chromosome must be copied. This occurs during the S phase of the cell cycle. After a chromosome is copied, the two duplicate strands of DNA (which are physically joined together by a structure located in the center called a *centromere*) are referred to as *chromatids*. With each of the 46 chromosomes having an attached duplicate, the cell can then begin the process of cell division. By aligning the chromatids in the center of the cell during metaphase, the cell ensures that each side of the cell will obtain one copy of each of the 46 chromosomes following cell division.

For more information on this concept, be sure to focus on:

- In Section 9.3, The DNA of each species is organized into a distinctive karyotype
- Figure 9.3, The Packing of DNA into a Chromosome
- Figure 9.4, Human Chromosomes
- Figure 9.5, The Stages of Cell Division

Meiosis

The key to understanding meiosis is knowing the overall goal of the process—to reduce the number of chromosomes in the daughter cells (gametes). Gametes in sexually reproducing species are haploid, that is, they contain only one copy of each chromosome. This process of producing haploid cells requires two cell divisions occurring in sequence, called *meiosis I* and *meiosis II*. The key step in the process is the pairing of homologous chromosomes, forming bivalents, during meiosis I. This pairing process does not occur during mitosis. When these bivalents are split during anaphase I, one complete chromosome (consisting of two identical chromatids) is segregated into each of the two daughter cells. The result is that each daughter cell formed after meiosis I has only one complete set of chromosomes, and therefore would be considered haploid. The process of meiosis II then serves to separate the chromatids into one of the two resulting daughter cells. In the end, the process of meiosis produces four gametes, each with one copy of each of the chromosomes.

For more information on this concept, be sure to focus on:

- In Section 9.4, Meiosis I is the reduction division
- In Section 9.4, Meiosis II is similar to mitosis
- Figure 9.8, Similarities and Differences between Meiosis and Mitosis

Tying It All Together

Several concepts presented in this chapter build upon concepts presented in previous chapters and may also be revisited and discussed in greater detail in subsequent chapters, including:

Cell Communication

- Chapter 6 – Section 6.6, Signaling Molecules in Cell Communication

Chromosomes and Genetics

- Chapter 11 – Section 11.1, The Role of Chromosomes in Inheritance

Crossing Over

- Chapter 11 – Section 11.3, Linkage and Crossing-Over

Practice Questions

Factual Knowledge

1. Which of the following important body functions does *not* typically involve some type of cell division?
   a. sexual reproduction
   b. repair of tissue damage
   c. making ATP from glucose
   d. development
   e. growth

2. Certain types of cell growth and repair can occur only with the involvement of a unique class of cells that can divide and remain unspecialized for the entire lifetime of an organism. These cells are known as
   a. gametes.
   b. muscle cells.
   c. skin cells.
   d. stem cells.
   e. cancer cells.

3. Before a cell can enter mitosis, it must first pass through
   a. interphase.
   b. G_1 phase.
   c. G_2 phase.
   d. S phase.
   e. all of the above

4. Which of the following statements about the G_0 phase is *true*?
   a. Cells in this phase show no evidence of having passed through the S phase.
   b. It is a phase that prepares the cell for mitosis.
   c. It is a phase that prepares the cell for meiosis.
   d. It is the phase in which DNA is replicated in preparation for cell division.
   e. all of the above

5. ________ is the process whereby an equal amount of parental genetic material is distributed between daughter nuclei destined for separate daughter cells.
   a. DNA duplication
   b. DNA segregation
   c. Meiosis
   d. Prometaphase
   e. Cytokinesis

6. A physician suspects that a chromosomal abnormality is the cause of a patient's disease. He orders a karyotype for this patient, which will allow the doctor to
   a. see the genes located on the patient's chromosomes.
   b. determine which chromosomes (maternal or paternal) are present in the patient's cells.
   c. see the number of mutations the patient has.
   d. count the total number of chromosomes present in the patient's cells.
   e. all of the above

7. In which of the following stages of mitosis would you expect to see chromatids become daughter chromosomes?
   a. prophase
   b. prometaphase
   c. metaphase
   d. anaphase
   e. telophase

8. The key feature of metaphase in mitosis is that
   a. the nuclear membrane begins to break down.
   b. new DNA is synthesized.
   c. chromosomes are no longer in a duplicated form.
   d. individual chromosomes line up along the equator of the cell.
   e. all of the above

9. Cytokinesis, the physical act of cell separation, occurs by the same mechanism in both plants and animals. (True or False)

10. The most important characteristic that distinguishes meiosis from mitosis is
    a. the total number of chromosomes present in daughter cells.
    b. the number of times that each stage of the division process occurs.
    c. whether or not the chromosomes are duplicated at the end of cell division.
    d. the presence of metaphase in mitosis, but not in meiosis.
    e. all of the above

11. The reduction division of meiosis specifically occurs at the end of meiosis II. (True or False)

12. The second division of meiosis is essentially
    a. the repeat of the first division of meiosis.
    b. a way to turn haploid cells into diploid cells.
    c. mitosis that begins with two haploid cells and ends with four haploid cells.
    d. mitosis that begins with two diploid cells and ends with four diploid cells.
    e. an unnecessary holdover from the early evolution of cells.

13. Which of the following would *not* be found in cells undergoing anaphase II of meiosis?
    a. daughter chromosomes
    b. homologous pairs of chromosomes
    c. DNA

d. a haploid number of chromosomes
e. centromeres

14. Match each term with the best description.
__ sister chromatids
__ G_0 phase
__ haploid
__ prophase
__ telophase
a. the product of DNA replication during S phase
b. stage of division in which cytokinesis occurs
c. resting stage for the cell in which no cell division occurs
d. immediately follows end of interphase
e. no homologous chromosomes present

Conceptual Understanding

1. The aftereffects of a sunburn, as depicted in Figure 9.1, show how skin cells are replaced as they are lost from the surface of the skin. This replacement process represents an example of
a. mitosis.
b. meiosis I.
c. meiosis II.
d. meiosis I and II.
e. none of the above

2. In Figure 9.2 in your textbook, what is the difference between the cell shown at the beginning of G_2 phase and the one shown entering mitosis and cytokinesis?
a. cell size only
b. cell size and chromosome number
c. cell size and whether or not chromosomes have been duplicated
d. chromosome number only
e. chromosome number and whether or not chromosomes have been duplicated

3 If you compare the chromosomes of a cell as it enters interphase with the same chromosomes just before the cell begins mitosis, a key difference would be
a. the total number of chromosomes that are present.
b. whether or not the chromosomes are duplicated.
c. whether or not homologous pairs of chromosomes are present.
d. the number of different kinds of chromosomes that are present.
e. all of the above

4. A cell whose haploid number is 12 will have ________ chromosomes present at prometaphase of mitosis.
a. 48
b. 24
c. 12
d. 6
e. cannot determine from the information given

5. In humans, mitosis
a. produces an exact genetic copy of the parent cell.
b. results in diploid daughter cells.
c. does not produce sperm or eggs.
d. makes daughter cells whose chromosomes are unduplicated.
e. all of the above

6. A normal diploid cell from a newly identified mammal has a total of 50 chromosomes. Which of the following statements about this cell is *true*?
a. If it undergoes meiosis, the daughter cells will have 50 chromosomes.
b. The cell contains 24 homologous pairs of chromosomes plus the sex chromosomes.
c. In metaphase of mitosis, this cell will have 50 chromatids per cell.
d. In metaphase II of meiosis, there will still be 50 chromosomes per cell.
e. all of the above

7. Consider the stage of meiosis called *metaphase II.* How many pairs of homologous chromosomes would be present in a cell undergoing metaphase II?
a. four
b. two
c. one
d. none
e. cannot determine from the information given

8. Consider a cell that begins meiosis with 30 total chromosomes. How many chromosomes will be present in each resulting daughter cell by the time the division process reaches metaphase II?
a. 0
b. 15
c. 20
d. 30
e. 60

9. If you randomly removed a cell from the human body, odds are that it would
a. have been produced by meiosis.
b. be haploid.
c. have 46 chromosomes.
d. have 23 chromosomes.
e. be in the part of the cell cycle called mitosis and cytokinesis.

10. The main difference between plant and animal cell division is that
a. cytokinesis takes place in different ways.
b. meiosis occurs only in animals.

c. mitosis occurs only in plants.
d. plants never produce diploid cells.
e. metaphase does not occur in plant cell division.

11. In which of the following stages would a human cell *not* be diploid?
    a. prophase of mitosis
    b. telophase of mitosis
    c. metaphase of mitosis
    d. prophase of meiosis I
    e. metaphase of meiosis II

12. You are asked to look at a microscope slide containing dividing human cells. Focusing on one of these cells, you note that it has 23 unduplicated chromosomes. This cell is most likely in which of the following stages of cell division?
    a. prometaphase of mitosis
    b. telophase of meiosis II
    c. anaphase of mitosis
    d. metaphase of meiosis I
    e. prophase of meiosis II

13. The painful arthritic symptoms caused by Lyme disease are related to cell division because
    a. joint cells are stuck in G_0 phase and cannot work properly.
    b. the tissues of the affected joints are produced by meiosis.
    c. the immune cells that attack healthy joint tissue result from mitosis.
    d. the Lyme bacterium stimulates mitotic production of faulty joint cells.
    e. all of the above

Related Activities

- Search your local newspaper or the World Wide Web to determine whether any cases of Lyme disease have been reported in your state or locality. Investigate what is known about how the disease-causing bacterium is transmitted from wildlife to people, how afflicted people can recognize that they have been exposed, and what the prescribed medical treatment is for Lyme infection. Compose a one-page summary highlighting the precautionary measures that are suggested for persons who spend time outdoors in areas where the Lyme disease bacterium is known to occur.
- Mitosis produces genetically identical cells, whereas meiosis leads to the production of genetically variable offspring. Most organisms use some form of sexual reproduction based upon meiosis. Propose reasons why having genetic variability in offspring might be advantageous over producing offspring that are genetic copies of the parent.
- As discussed in the newspaper article "It's Time to Play the Music, It's Time to Light the Lights" in the Biology in the News section at the end of the chapter, prokaryotes are highly efficient at undergoing cell division. The pneumonia-causing organism *Streptococcus pyogenes* is considered one of the class of "flesh-eating" bacteria. Search the World Wide Web to identify at least one other "flesh-eating" bacteria and compose a one-page summary discussing the effects of contamination caused by your identified microbe and the recommended treatment regimen.

Answers and Explanations

Factual Knowledge

1. **c** All the other choices involve the addition of new cells to a body or the formation of a new organism; they thus require some type of cell division. ATP formation occurs as a result of biochemical activities within an existing cell.
2. **d** Refer to Figure 9.1 to see how stem cells are used to replenish skin cells as they are lost.
3. **e** The entirety of interphase (G_1, G_2, and S) consists of events that are necessary precursors to mitosis.
4. **a** G_0 is a stasis in which cells are temporarily or permanently prevented from dividing. When stuck in G_0, cells cannot complete the growth and synthesis portions of interphase, and thus are unable to divide. Most cells of the body are at least temporarily stuck in G_0 once adulthood is reached.
5. **b** "Segregation" means to separate. DNA segregation is the equal separation of chromosomal material into daughter cells. This stage occurs during mitosis.
6. **d** Recall that a karyotype is a pictorial representation of the chromosomes taken from a cell about to undergo division.
7. **d** Sister chromatids separate and become daughter chromosomes during anaphase of mitosis. Refer to Figure 9.5 in your textbook to see a diagram of this process.
8. **d** See Figure 9.5 in your textbook; the single-file arrangement of chromosomes along the metaphase plate characterizes metaphase.
9. **b** Since plant cells have a cell wall, cytokinesis cannot occur through the simple inward pinching of

membranes, as it does in animal cells. In plants, a new cell wall is first laid down between what will become the daughter cells, and then a membrane forms to complete the separation process. Refer to Figure 9.6 in your textbook to see a diagram of this process.

10. **a** Recall that the overall purpose of mitosis is to make genetically identical daughter cells (diploid parent cell to diploid daughter cells), while the purpose of meiosis is to reduce the number of chromosomes by one-half (diploid parent cell to haploid daughter cells).

11. **False** The reduction division that turns a diploid cell into haploid daughter cells occurs in meiosis I.

12. **c** The purpose of meiosis II is to separate the attached duplicated sister chromatids present in the haploid cells following meiosis I. This is essentially what mitosis does, only with diploid cells instead of haploid.

13. **b** Recall that homologous pairs of chromosomes are separated during meiosis I, and therefore would not be present at anaphase II.

14. **a** sister chromatids
    c G_0 phase
    e haploid
    d prophase
    b telophase

Conceptual Understanding

1. **a** This is an example of producing new body cells, which is accomplished through mitosis.

2. **a** No changes in the amount of DNA or chromosome number occur after S phase. The only thing that changes is cell size.

3. **b** The primary purpose of the S phase, which is a major portion of interphase, is to duplicate chromosomes in preparation for cell division.

4. **b** Cells undergoing mitosis are diploid. If the haploid number of a cell is 12, then the diploid number is 24.

5. **e** All of these statements are true of mitosis in humans and in any diploid organism.

6. **b** A diploid cell with 50 chromosomes has 25 pairs, one of which, for most multicellular organisms, are the sex chromosomes. During mitosis the cell will remain diploid, so in metaphase there will be 100 chromatids. Meiosis in this cell will produce haploid daughter cells with only 25 chromosomes.

7. **d** The first division of meiosis (reduction division) yields haploid cells, so throughout all of meiosis II there are no homologous pairs of chromosomes present in the cells.

8. **b** By the time the cells reach metaphase II, they have already completed meiosis I (reduction division) and are therefore haploid. Each cell will have 15 chromosomes.

9. **c** Randomly removing a cell from a human would likely give you a body cell, which would be diploid and have 46 chromosomes. It would also likely be in interphase, since that is the longest part of the cell cycle.

10. **a** Both mitosis and meiosis have similar division events in plants and animals. The main difference is cytokinesis, which must be different due to the presence of the plant cell wall. Refer to Figure 9.6 for a diagram of this process.

11. **e** Cells are diploid throughout all of mitosis and throughout most of the first division of meiosis.

12. **b** The diploid number for humans is 46, so this cell must be haploid and have already undergone meiosis I. Given that its chromosomes are unduplicated, it must have completed anaphase of meiosis II when the duplicate chromatids are separated. That leaves telophase II as the most likely stage.

13. **c** New immune cells are produced through mitosis. Lyme disease causes an overproduction of immune cells that then mistakenly attack the joint tissues of the body thinking they are attacking bacterial cells instead. This results in the painful arthritic symptoms associated with the disease.

INTERLUDE **B**

Cancer: Cell Division Out of Control

Getting Started

Below are a few questions to consider prior to reading Interlude B. These questions will help guide your exploration and assist you in identifying some of the key concepts presented in this interlude.

1. How have scientists used adenoviruses as a tiny anticancer weapon?
2. How do positive and negative growth regulators work together to ensure cells do not continue to grow in an unregulated fashion?
3. Who was Peyton Rous, and what was his contribution to the study of cancer?
4. How are proto-oncogenes and tumor suppressor genes analogous to two ponies pulling a cart in opposite directions?
5. What percentage of American males will develop invasive cancer over their lifetime?
6. What percentage of all human cancers involve the complete loss of the tumor suppressor gene *p53*?

A Guide to the Reading

The following concepts typically give students the most difficulty when exploring the content in Interlude B for the first time. For each concept, one or more references have been identified that may help you gain a better understanding of these potentially problematic areas.

Proto-oncogenes and Tumor Suppressors

As described in the Interlude, genes responsible for causing cancer are referred to as *oncogenes*. Oncogenes typically are genes that have a normal cellular function, but have become mutated or over-activated. When this occurs, the altered gene then serves to stimulate cell division, resulting in unregulated cell growth and the development of cancer. Because these genes can exist in two states, they are often called *proto-oncogenes* (normal) or *oncogenes* (altered, cancer-causing). Luckily, the cell does not rely on the health of proto-oncogenes to avoid the development of cancer. Typically, there are several safeguards that help to control the cell division process. These safeguards are called *tumor suppressors*—genes whose function it is to halt cell (and tumor) growth. Therefore, it is the balance between the action of proto-oncogenes and tumor suppressors that controls whether a cell continues to divide. When this system fails, the result is tumor formation.

For more information on this concept, be sure to focus on:

- In Interlude B, Oncogenes play an important role in cancer development
- In Interlude B, Negative Growth Regulators: Inhibiting Cell Division
- Figure B.3, The Rous Sarcoma Virus Causes Cancer in Chickens
- Figure B.7, The Control of Cell Division by Proto-oncogenes and Tumor Suppressor Genes

Cancer as a Multi-Step Process

The development of tumors can result when the cell division regulatory machinery becomes defective. As described in the Interlude, this regulatory machinery involves a complex series of safeguards that help to protect the cell. In order for a cell to become cancerous, several of these safeguard checkpoints must be overcome. A particularly good example described in the chapter is that of colon cancer. In order for a malignant tumor to develop in the colon, a cell in the wall of the colon must first lose the function of two or more tumor suppressor genes, have a mutation in a proto-oncogene that results in the gene becoming an oncogene, and completely lose the *p53* tumor suppressor gene. If all of these events occur in the proper sequence, a small benign polyp may ultimately develop into a malignant tumor. The key to understanding this process is realizing that if any of these conditions are not met, cancer will not develop.

For more information on this concept, be sure to focus on:

- In Interlude B, An Interplay of Factors Can Cause Cancer
- In Interlude B, Cancer is a multi-step process
- In Interlude B, Cancer is often related to lifestyle choices
- Figure B.8, Colon Cancer Is a Multi-step Process

Tying It All Together

Several concepts presented in this chapter build upon concepts presented in previous chapters and may also be revisited and discussed in greater detail in subsequent chapters, including:

Viruses

- Chapter 3 – Section 3.7, The Difficulty of Viruses

Cell Communication and Signaling

- Chapter 6 – Section 6.6, Signaling Molecules in Cell Communication

Chromosomes and Karyotype

- Chapter 9 – in Section 9.3, The DNA of each species is organized to give it a distinctive karyotype

Human Genetics

- Chapter 11 – Section 11.1, The Role of Chromosomes in Inheritance

Mutation

- Chapter 12 – Section 12.4, Repairing Replication Errors and Damaged DNA

Gene Therapy

- Chapter 15 – Section 15.3, Applications of DNA Technology

Practice Questions

Factual Knowledge

1. The suggestion that modified adenoviruses might serve as a treatment for cancer comes from an understanding of
   a. how viruses target particular cells.
   b. how viruses take control of cells.
   c. tumor suppressor gene function.
   d. the role of *p53* in controlling cell division.
   e. all of the above

2. Which of the following is a type of cancer *not* found in humans?
   a. colon cancer
   b. Rous sarcoma
   c. retinoblastoma
   d. lung cancer
   e. skin cancer

3. Which of the following occurs when the Rous sarcoma virus attacks healthy cells?
   a. Viral oncogenes are turned off.
   b. Protein kinase activity declines.
   c. An oncogene activates cell division.
   d. A tumor suppressor gene is deactivated.
   e. all of the above

4. The *Src* gene found in the Rous sarcoma virus
   a. interferes with the negative control mechanisms of cell division.
   b. increases the normal effect of proto-oncogenes in the host cell.
   c. is a mutation that the virus acquires from its most recent host cell.
   d. results in the production of an overactive protein kinase.
   e. none of the above

5. Most human proto-oncogenes are
   a. very uncommon and almost never mutate into oncogenes.
   b. common but unlikely ever to be changed into oncogenes.
   c. common and changed into oncogenes mainly by viral infections.
   d. common and changed into oncogenes mainly by environmental factors.
   e. always changed into the *Src* gene by viral infection.

6. One reason proto-oncogenes are especially susceptible to becoming oncogenes is that they code for the production of _______ that are part of the normal signal cascades involved in the control of cell division.
   a. growth factors
   b. *Src* genes
   c. tumor suppressors
   d. phospholipids
   e. all of the above

7. The eye cancer known as *retinoblastoma* results from
   a. the reduction or absence of a tumor suppressor gene product.
   b. increased activity by tumor suppressor genes.
   c. the addition of genetic material to chromosome 13.
   d. excessive production of the Rb protein.
   e. all of the above

8. Unlike oncogenes, both copies of a tumor suppressor gene must be mutated for a cell to lose control of its negative growth regulation. (True or False)

9. If a cell has complete control over both negative and positive regulation of its division, which of the following will most likely occur?
   a. The cell will divide rapidly but not become a tumor.
   b. The cell will divide to become a benign tumor.
   c. The cell will divide to become a malignant tumor.
   d. The cell will divide at its normal rate.
   e. No signal cascade molecules will be produced in the cell.

10. Nearly half of all cancers can be traced to purely inherited genetic defects. (True or False)

11. Imagine that a friend of yours has a small polyp removed from his colon. This probably means that he has
    a. incurable colon cancer.
    b. a malignant but treatable form of colon cancer.
    c. a benign form of colon cancer that has been discovered at an early stage.
    d. acquired a cancer-causing virus.
    e. a cancer that has spread to other organs.

12. In which of the following human organs do 100,000 or more new cases of cancer develop annually in the United States?
    a. breast
    b. lung
    c. prostate
    d. colon
    e. all of the above

13. Which of the following statements regarding cancer risk is *true*?
    a. Identical twins have a higher risk of getting cancer than nonidentical twins.
    b. Inherited genetic defects pose the most significant risk to developing cancer.
    c. Environmental factors pose the greatest cancer risk.
    d. Viruses are the only known risk to developing cancer.
    e. none of the above

14. Smoking has been linked to which of the following cancers?
    a. lung
    b. bladder
    c. kidney
    d. stomach
    e. all of the above

15. Match each term with the best description.
    __ benign
    __ malignant
    __ *p53*
    __ proto-oncogene
    __ tumor suppressor
    a. an example of a critical tumor suppressor gene
    b. tumor that spreads to other parts of the body
    c. Rb protein is an example
    d. relatively harmless tumor
    e. predecessor of oncogene

Conceptual Understanding

1. Cigarette smoking has been shown to cause cancer. Which of the following statements regarding the link between smoking and cancer is *false*?
    a. Cigarette smoke contains carcinogens such as polycyclic aromatic hydrocarbons (PAHs).
    b. PAHs form physical complexes with DNA called adducts.
    c. The *p53* gene contains several sites for adduct formation.
    d. The formation of adducts in the *p53* gene is restricted to lung cells in smokers.
    e. none of the above

2. Refer to Figure B.2 in your textbook. In Peyton Rous's landmark experiment, which of the following conclusions could be made?
    a. The filtrate contained a cancer-causing agent.
    b. The cancer-causing agent contained in the filtrate could not have been due to a bacteria.
    c. The cancer-causing agent contained in the filtrate had to be a virus.
    d. The filtrate would consistently cause cancer when injected into new chickens.
    e. none of the above

3. What do p53 and Rb have in common?
    a. They are both oncogenes.
    b. They are both proto-oncogenes.
    c. They are both tumor suppressor proteins.
    d. They both promote excessive cell division.
    e. Neither is involved in regulating signal cascades.

4. For a cell to undergo the process of normal cell division,
    a. proto-oncogenes must be activated and tumor suppressors must be inactivated.
    b. proto-oncogenes must be inactivated and tumor suppressors must be activated.
    c. oncogenes must be activated and proto-oncogenes must be inactivated.
    d. oncogenes must be activated and tumor suppressors must be inactivated.
    e. none of the above

5. In human retinal cells, an absence of the Rb protein will
    a. have no effect on cell division.
    b. increase cell division.
    c. decrease cell division.
    d. allow signal cascades to proceed normally.
    e. cause oncogenes to revert back to proto-oncogenes.

6. The child shown in Figure B.5 of your textbook has retinoblastoma. Which of the following is true about this child's condition?
    a. Both copies of the *Rb* gene are defective in tumor cells from the affected eye.
    b. One copy of the *Rb* gene is defective in tumor cells from the affected eye.
    c. Neither copy of the *Rb* gene is defective in tumor cells from the affected eye.
    d. Both copies of the *Rb* gene are defective in all retinal cells from both eyes.
    e. This child probably has an inherited case of retinoblastoma.

7. Consider the middle panel in Figure B.7 of your textbook. What would be the effect on the horse-and-cart analogy if the positive regulation proto-oncogenes on both chromosomes were mutated into oncogenes?
   a. Both horses would disappear.
   b. The pony on the left would also become a draft horse.
   c. The draft horse on the right would become two draft horses.
   d. The horse on the right would disappear.
   e. The cart would disappear.

8. The vast majority of human cancer cases
   a. are colon cancer.
   b. occur in children.
   c. involve women.
   d. are caused by inherited genetic conditions.
   e. are caused in part or completely by environmental factors.

9. It usually takes only one or two mutations of proto-oncogenes to trigger the development of most cancers in humans. (True or False)

10. By the time that human colon cancer has developed to the malignant carcinoma stage, at least ________ different mutations have occurred in the regulatory controls associated with normal cell division.
    a. 2
    b. 4
    c. 8
    d. 20
    e. 50

11. The modified adenoviruses currently being tested as possible anticancer treatments work by attacking cells that
    a. have normal proto-oncogenes.
    b. have plenty of Rb tumor suppressor proteins.
    c. have completely lost their p53 tumor suppressor protein activity.
    d. are infected with the Rous sarcoma virus.
    e. show evidence of colon cancer polyps.

12. In colon cancer, the complete loss of one tumor suppressor, combined with the mutation of a proto-oncogene, leads to the formation of
    a. benign polyps.
    b. middle-stage polyps.
    c. late-stage polyps.
    d. a malignant tumor.
    e. none of the above

13. Both colon cancer and smoking-induced lung cancer
    a. are inherited genetic defects.
    b. are caused by an overactive proto-oncogene.
    c. involve inactivation of *p53*.
    d. are caused by rare viral infections.
    e. none of the above

Related Activities

- Search the World Wide Web for information on the relation of diet to colon cancer. What kinds of food are thought to increase the occurrence of colon cancer, and which to reduce it? What is known about the way diet affects the positive and negative growth regulation factors that control cell division? Compose a one-page essay summarizing the information you find.
- Search the World Wide Web for additional information about the human cancers listed in Table B.1 of your textbook. Discuss any environmental factors that are thought to be associated with the development of these cancers, and how these factors specifically cause the mutation of proto-oncogenes into oncogenes. Compose a one-page essay summarizing the results of your search.
- What is your opinion of the research presented in the Biology Matters article, "The Truth about Cigarettes in School" at the end of this Interlude, which discusses the link between exposure to benzopyrene and the development of cancer. The article discusses the differences between cigarettes and marijuana. In your opinion, which of these poses a greater cancer risk to college students? Explain your answer.

Answers and Explanations

Factual Knowledge

1. **e** Understanding the way that viruses seek out their host and alter their division, combined with a knowledge of how the loss of tumor suppression control in cancer cells is related to the function of *p53*, has opened the door to what may prove to be an effective method of selectively destroying tumor cells using modified adenoviruses.
2. **b** This type of tumor is found in chickens.
3. **c** The insertion of the *Src* viral oncogene into the host cell results in the production of an overactive protein kinase that stimulates cell division.
4. **d** The *Src* oncogene has its effect on positive regulation of cell division by stimulating production of an overactive protein kinase. This mutation probably arose in the distant history of the virus and does not alter proto-oncogenes in the host cell.

5. **d** Although some human cancers may be triggered by viruses, most are the result of numerous proto-oncogenes being mutated into oncogenes by environmental factors such as chemical pollutants.

6. **e** Signaling cascades receive signals from outside of the cell and work to stimulate various proteins, including growth factors, inside of the cell through a series of stepwise activation steps.

7. **a** Retinoblastoma develops when the Rb tumor suppressor protein (a part of the negative growth regulation mechanism) is absent from a cell, sometimes as a result of the loss of genetic material from chromosome 13. Even when the *Rb* gene is present in the cell, it has decreased activity.

8. **True** Since the protein product made by a tumor suppressor gene has a negative influence on cell division, both copies of such a gene must be inactive to completely eliminate the negative control on cell division.

9. **d** This set of circumstances provides the regulatory balance that will lead to normal cell division.

10. **False** In very rare cases, oncogenes are inherited. Rather, environmental factors play a huge role in determining cancer risk, often in combination with genetic factors.

11. **c** A small polyp in the colon is usually a sign that the tumor is in its earliest stages of development, and is therefore likely to be benign. Refer to Figure B.8 for additional information.

12. **e** See the information given in Table B.1 of your textbook.

13. **c** As demonstrated by the Scandinavian twin study, the most significant factor for cancer risk was the environment.

14. **e** Smoking (along with human behavior), recognized as the leading cause of cancer in the United States, has been linked to all of these malignancies.

15. **d** benign
b malignant
a p53
e proto-oncogene
c tumor suppressor

Conceptual Understanding

1. **d** Recall that the formation of adducts in the p53 gene due to exposure to PAHs is not restricted to lung cells. White blood cells have also shown similar susceptibility that can lead to other forms of cancer, such as leukemia.

2. **c** While Rous's experiments led to the discovery of the Rous sarcoma virus, his experiments did not definitely demonstrate that a virus was involved in the process.

3. **c** Both p53 and Rb are tumor suppressor proteins that play a role in the development of cancer.

4. **a** In other words, the promoting effects of positive controls must be "on" while the inhibitory effects of negative controls must be "off" to get normal cell division.

5. **b** The absence of Rb protein means that an important negative control on cell division is inactivated. This greatly increases the chance of tumor development.

6. **a** Since Rb is a tumor suppressor protein, both genes that code for this substance must be defective in the retinoblastoma tumor cells.

7. **c** This type of mutation would further promote cell division, which would be analogous to having a second strong draft horse pulling against the negative regulation of the pony. Tumor development would thus be even more likely.

8. **e** Environmental factors play at least some role in over 95 percent of human cancers.

9. **False** Usually human cancers develop as the result of a series of mutations involving several proto-oncogenes.

10. **b** See Figure B.8 in your textbook for this information.

11. **c** Normal cells still have p53 protein activity, and so will not be attacked by the modified virus. Only cancer cells have lost all p53 protein function, which is one reason they divide uncontrollably.

12. **b** Again, see Figure B.8 in your textbook for this information.

13. **c** The p53 protein plays a role in both of these types of cancer. In colon cancer, inactivation of *p53*, in conjunction with the loss of a tumor suppressor and over-activation of a proto-oncogene, can result in the formation of carcinomas. In the case of lung cancer, repeated exposure to the carcinogens present in cigarette smoke eventually inactivates the *p53* present in lung cells, leading to the development of tumors.

CHAPTER 10 Patterns of Inheritance

Getting Started

Below are a few questions to consider prior to reading Chapter 10. These questions will help guide your exploration and assist you in identifying some of the key concepts presented in this chapter.

1. What was the legend of Princess Anastasia, and how did the science of genetics play a part in solving this mystery?
2. What organism did the "Father of Genetics," Gregor Mendel, work with to devise his theory of inheritance?
3. What is the ultimate source of all new genetically inherited traits?
4. Why do true-breeding purple-flowered pea plants, when crossed with true-breeding white-flowered pea plants, only produce purple-flowered offspring?
5. How did the experiments of Gregor Mendel disprove the blending theory of inheritance?
6. What is a Punnett square and how can it be used to illustrate Mendel's law of segregation?
7. What is incomplete dominance?
8. Why might a mouse with two copies of the dominant black fur gene still have white fur?
9. What are polygenic traits, and how do they determine characteristics such as skin color in humans?

A Guide to the Reading

The following concepts typically give students the most difficulty when exploring the content in Chapter 10 for the first time. For each concept, one or more references have been identified that may help you gain a better understanding of these potentially problematic areas.

Dominant vs. Recessive

The concept of gene dominance is vital to the understanding of inheritance. Recall that the phenotype of an organism refers to the physical, observable trait resulting from the individual's genotype, or genetic makeup. A perfect example of this is flower color in pea plants as studied by Mendel. In this case, flower color is determined by two different alleles, or versions, of a single gene. The *P* allele contains the genetic code for producing purple flowers (the phenotype). The *p* allele contains the genetic code for producing white flowers. When a single copy of each allele is present in the same plant, the resulting genotype of the organism is *Pp*. Since the two copies of the gene are dissimilar, we refer to this organism as a *heterozygote*. In heterozygous organisms, the influence of one gene may take precedence over the other in the production of the phenotype. In this example, the heterozygous plant would produce purple flowers. Therefore, the *P* allele is considered to be dominant to the recessive *p* allele since the purple flower phenotype is displayed over the white flower phenotype. Keep in mind that the terms *dominant* and *recessive* are comparative; an allele, if considered on its own, cannot be determined to be dominant or recessive unless one is comparing it to an alternate form of the same gene.

For more information on this concept, be sure to focus on:

- Section 10.1, Essential Terms in Genetics
- Section 10.3, Basic Patterns of Inheritance

Mendel's Law of Segregation

As stated in the chapter, the law of segregation states that "the two copies of a gene separate during meiosis and end up in different gametes." To understand this statement, be sure to review the discussion of meiosis in Chapter 9 (in particular Section 9.4, Meiosis: Halving the Chromosome Number). If you understand that it is physically impossible for more than one copy of a single gene to end up in a single sperm or egg cell (under normal circumstances), then understanding the law of segregation becomes easy. Keep in mind that it is this law of segregation that makes it possible to utilize the Punnett square to predict the probability of obtaining offspring with a given genotype and phenotype.

For more information on this concept, be sure to focus on:

- In Section 10.4, Mendel's first law: Segregation
- Figure 10.6, The Punnet Square Method

Mendel's Law of Independent Assortment

As stated in the text, the law of independent assortment states that "when gametes form, the separation of the two copies of one gene during meiosis is independent of the separation of the copies of other genes." This law basically states that each gene present in an organism is inherited independently of all of the other genes. Mendel showed this with his experiments on pea plants while observing the inheritance of multiple traits (such as pea color and shape) simultaneously. Recall that while the pea color trait (yellow versus green) and the pea shape trait (wrinkled versus smooth) both obey the law of segregation as expected, these two traits behave independently of each other during the process of meiosis. Therefore, the inheritance of a particular pea color has no influence whatsoever on the inheritance of pea shape.

For more information on this concept, be sure to focus on:

- In Section 10.4, Mendel's second law: Independent assortment
- Figure 10.7, Independent Assortment of Alleles
- Figure 10.8, Are Genes Inherited Independently?

Incomplete Dominance and Codominance

Not all traits follow a simple dominant/recessive pattern. For these traits, an allele may not produce its maximum phenotype or may contribute to multiple phenotypes. Consider a gene with two alleles that produce distinct phenotypes. When heterozygotes display a phenotype that is intermediate between the two phenotypes displayed by homozygotes, this is referred to as *incomplete dominance.* The examples used in the text include flower color in snapdragons and coat color in horses. The key to identifying incomplete dominance is the fact that heterozygous individuals display a different phenotype than either of the two homozygous forms. This is in contrast to *codominance,* where "the phenotype of the heterozygote is determined equally by each allele." In this case, different alleles of the same gene may produce their maximum phenotype simultaneously. The example provided in the text is that of blood types in humans, where two alleles (A and B) will be expressed regardless of the combination of alleles inherited. The key to identifying codominance is understanding that heterozygous individuals essentially display two phenotypes (one for each different allele) simultaneously. In the case of blood types, an individual who is heterozygous (AB) will display the type A phenotype *and* the type B phenotype simultaneously. Blood types are complicated slightly by the presence of a third allele (O), which is recessive to both the A and the B alleles.

For more information on this concept, be sure to focus on:

- In Section 10.5, Many alleles do not show complete dominance
- Figure 10.9, Incomplete Dominance in Horses
- Figure 10.10, Codominance in the Blood Typing System

Epistasis

Genes are complex entities. Rarely does a gene exert its phenotypic effect without interaction from the presence or absence of other genes. When the effect of a particular allele depends on the presence of other alleles from a separate, distinct gene, this is called *epistasis.* The example provided in the text for epistasis is coat color in mice. Separate genes for the pigment melanin and a biochemical precursor to the pigment are inherited independently. In this case, the alleles for melanin production result in either black fur (*B*, dominant) or brown fur (*b*, recessive). Both alleles for pigment production rely on the second gene for the precursor, which has two alleles as well (*C*, dominant, pigment produced; *c*, recessive, no pigment produced). Therefore, regardless of the genotype for fur color, if the mouse is homozygous recessive for the precursor (*cc*), the mouse will be incapable of producing pigment, and will be white. In this case, the melanin gene is dependent upon the independently inherited precursor gene. When considering epistasis, it is important to remember that even though multiple genes may interact to produce a single phenotype, these genes still must obey the law of independent assortment.

For more information on this concept, be sure to focus on:

- In Section 10.5, Alleles for one gene can alter the effects of another gene
- Figure 10.11, Gene Interactions

Tying It All Together

Several concepts presented in this chapter build upon concepts presented in previous chapters and may also be revisited and discussed in greater detail in subsequent chapters, including:

Sex-linked Inheritance

- Chapter 11 – Section 11.7, Sex-Linked Inheritance of Single-Gene Mutations

Exceptions to Mendelian Genetics

- Chapter 11 – Section 11.3, Linkage and Crossing-Over

Mutations

- Chapter 13 – Section 13.6, The Effect of Mutations on Protein Synthesis

DNA Technology and Human Disease

- Chapter 15 – Section 15.3, Applications of DNA Technology

Practice Questions

Factual Knowledge

1. Which of the following statements regarding genes is *false*?
   a. Genes are located on chromosomes.
   b. Genes consist of a long sequence of DNA.
   c. Genes contain information for the production of a single protein.
   d. In sexually reproducing species, each cell contains a single copy of every gene.
   e. none of the above

2. Mendel used ________ as his research organism to study patterns of genetic inheritance.
   a. garden peas
   b. snapdragons
   c. horses
   d. Siamese cats
   e. mice

3. Mendel's primary contribution to our understanding of genetic inheritance was
   a. the idea that genes are found on chromosomes.
   b. providing a mechanism that explains patterns of inheritance.
   c. describing how genes are influenced by the environment.
   d. determining that information contained in DNA codes for proteins.
   e. the discovery of alleles.

4. In Mendel's model of inheritance, what he referred to as *particles* we now refer to as
   a. genes.
   b. chromosomes.
   c. homozygotes.
   d. heterozygotes.
   e. phenotypes.

5. A diploid organism that is heterozygous for a given gene has the same number of alleles as does an organism that is homozygous for that gene. (True or False)

6. A purple-flowered pea plant has the genotype *PP*. Which of the following statements about this plant is *false*?
   a. Its phenotype is white.
   b. It has a homozygous dominant genotype.
   c. When bred to a white-flowered plant, its offspring will all be purple.
   d. The gametes produced will all have the same flower color allele.
   e. It is of the true-breeding variety.

7. In a parental cross between a true-breeding, purple-flowered pea plant and a white-flowered plant, the F_2 generation will
   a. all be purple.
   b. all be white.
   c. show a 3:1 ratio of purple to white.
   d. show a 3:1 ratio of white to purple.
   e. all be heterozygotes.

8. The Punnett square method is useful as a tool for analyzing genetic crosses because alleles separate equally during gamete formation. (True or False)

9. Being able to use a Punnett square to track the pattern of inheritance in a two-characteristic cross (e.g., pea seed color and seed shape) demonstrates Mendel's idea of

a. allele segregation.
b. blending inheritance.
c. environmental influences on genes.
d. independent assortment.
e. incomplete dominance.

10. Mendel's laws of segregation and independent assortment both have their biological basis in events that take place during mitosis. (True or False)

11. A true-breeding, purple-flowered pea plant is crossed with a white-flowered plant. They produce two F_1 progeny that are then mated with each other to yield ten F_2 plants. The F_2 generation contains five purple-flowered plants and five white-flowered plants. Which of the following is the best explanation for the F_2 results?
    a. They are exactly what is expected from Mendel's laws.
    b. They are similar to, but not exactly, what Mendel's laws predict.
    c. They are rather different from predictions, probably due to small sample size.
    d. The F_1 plants must have suffered a mutation before mating.
    e. None of the above is a likely explanation for these results.

12. A chestnut-colored horse is mated with a cremello-colored horse. Over a 10-year period, all of their offspring are palominos. This pattern of inheritance is best explained by
    a. complete dominance.
    b. incomplete dominance.
    c. multiple gene effects.
    d. environmental effects on genes.
    e. small sample sizes.

13. In humans, the genetic commonality among heart disease, cancer, alcoholism, and diabetes is that they are all
    a. caused by mutations of the same gene.
    b. strictly environmentally induced with little or no genetic component.
    c. clear violations of Mendel's basic laws of genetic inheritance.
    d. caused by multiple genes with a strong environmental influence.
    e. cases of genes exhibiting incomplete dominance.

14. Match each term with the best description.
    __ blending inheritance
    __ heterozygote
    __ recessive allele
    __ incomplete dominance
    __ phenotype

    a. observable characteristic of genes
    b. genotype with dissimilar alleles
    c. produces a heterozygote that is an intermediate in phenotype between the two homozogotes for a particular allele
    d. does not have phenotypic effect when paired with dominant allele
    e. popular theory prior to Mendel's research

Conceptual Understanding

1. Which of the following statements about mutations is *false*?
    a. Mutations are the source of new alleles for genes.
    b. Organisms are able to create mutations to meet their specific needs.
    c. Mutations are random events and can happen in any cell at any time.
    d. Most mutations tend to be harmful or have no effect on organisms.
    e. Occasionally, mutations can be beneficial to organisms.

2. Two organisms that are true-breeding for a certain genetic characteristic are mated and their offspring analyzed. Which of the following statements about this situation is *true*?
    a. Both parents are homozygotes.
    b. The offspring are either 100 percent homozygotes or 100 percent heterozygotes.
    c. The offspring represent the F_1 generation.
    d. The gametes produced by the offspring will carry only one allele for this gene.
    e. all of the above

3. A pea plant that is heterozygous for the flower gene makes gametes. What is the probability that one of its gametes contains the recessive white allele for flower color?
    a. 0%
    b. 25%
    c. 50%
    d. 75%
    e. 100%

4. The only known violation of Mendel's law of independent assortment is flower color in snapdragons, which is best explained using the concept of blending inheritance. (True or False)

5. The variation in coat color for the three mice pictured in Figure 10.11 of your textbook is the result of a single-gene trait modified by environmental influences. (True or False)

6. Consider a gene with two alleles that show complete dominance. When two heterozygotes for this gene breed, they have a 25 percent chance of producing a homozygous recessive offspring. The next time these two individuals breed, what are the chances that they will once again have a homozygous recessive progeny?
   a. 0%
   b. 25%
   c. 50%
   d. 75%
   e. 100%

7. Recall your textbook's explanation of probability using the number of times a coin is tossed. Now imagine fertilization events involving sperm and eggs instead of coin tosses. Which of the following number of fertilizations would most likely give you the predicted phenotypic ratio of offspring for some genetic cross?
   a. 4
   b. 10
   c. 20
   d. 100
   e. 10,000

8. Consider Figure 10.8 in your textbook. The 9:3:3:1 ratio of phenotypes that occurs in the F_2 generation of this cross can be explained using
   a. blending inheritance.
   b. Mendel's law of segregation.
   c. Mendel's law of independent assortment.
   d. both of Mendel's laws of genetic inheritance.
   e. incomplete dominance.

9. Genetically identical plant clones can exhibit dramatic phenotypic variation depending on the environmental conditions under which they are grown. (True or False)

10. Theoretically, a Siamese cat raised from birth in a very cold environment should have almost no black pigment in its fur. (True or False)

11. Recall the genetic determination of coat color in mice, as discussed in your textbook. A cross between a male with the genotype *BBcc* and a female with the genotype will produce offspring with which coat color(s)?
   a. black only
   b. brown only
   c. white only
   d. black and brown
   e. black, brown, and white

12. In a particular plant, two genes control leaf shape and color. Round leaves (*R*) are dominant to jagged leaves (*r*). Yellow fruits (*Y*) are dominant to white fruits (*y*). A true-breeding round-leaved, yellow-fruited plant is mated with a jagged-leaved, white-fruited plant. What are the genotypes of the plants involved in this cross?
   a. *RRYY* × *RRYY*
   b. *RRYY* × *rryy*
   c. *RrYy* × *RrYy*
   d. *RrYy* × *rryy*
   e. cannot determine from the information given

13. Using the information given in question 12, what is (are) the phenotype(s) of the offspring produced by this cross?
   a. round leaves, yellow fruit
   b. round leaves, white fruit
   c. jagged leaves, yellow fruit
   d. jagged leaves, white fruit
   e. cannot determine from the information given

14. Again using the information given in question 12, what is the probability that this cross will produce any offspring with the genotype *rrYy*?
   a. 0%
   b. 25%
   c. 50%
   d. 75%
   e. 100%

NOTE: For more practice with the concepts of genetic inheritance, we strongly suggest that you work through the sample genetics problems at the end of Chapter 10 in your textbook.

Related Activities

- Search the World Wide Web for three examples of the way basic mechanisms of genetic inheritance are being used to (1) produce better food, (2) breed better plants or animals for the gardening and dairy industries, and (3) fight human disease. Write a paragraph summarizing each example, and state how Mendelian principles of genetics apply.
- Many purebred dogs are particularly susceptible to genetic abnormalities. Research the topic of inbreeding in your library or on the Web, and find an example of genetic inbreeding in dogs. Explain how this inbreeding has occurred, what the biological consequences are for the dogs themselves, and what can be done to reduce the problem.
- Review the Biology in the News article at the end of the chapter. This article discusses how genes may influence animal behavior. In this example, the gene that is responsible for producing a receptor for the hormone vasopressin

is altered to manipulate sexual behavior in male voles. Search the World Wide Web to find additional information about the hormone vasopressin. Compose a one-page summary of some of the vasopressin irregularities that occur in humans and the behavioral effects these irregularities have. Considering these diseases, discuss how research on altering the vasopressin gene may potentially help individuals suffering from the identified conditions.

Answers and Explanations

Factual Knowledge

1. **d** Recall that in sexually reproducing species, each cell (except for gametes) contains two copies of every gene, one copy inherited from each parent.
2. **a** Mendel first used the garden pea to make his revolutionary study of inheritance patterns.
3. **b** Mendel gave us a viable mechanism for explaining inheritance. All of the other contributions were made by scientists who did their work after Mendel.
4. **a** We now know that genes (or alleles) are the "particles" that Mendel was referring to when he imagined their segregation into gametes.
5. **True** Since both organisms are diploid, both individuals would have the same number of alleles, just different combinations of alleles.
6. **a** According to the shorthand notation used in the textbook, *PP* is the homozygous dominant genotype that results in a purple phenotype, not white.
7. **c** In this cross, the F_1 generation would all be heterozygous for flower color (*Pp*), resulting in 100 percent purple-flowered plants, while the F_2 generation would show the expected 3:1 ratio.
8. **True** The first step in making a Punnett square is to assign the possible gametes that can be made by each parent to adjacent sides of the square, assuming that all gametes can be made with equal probability. This assumption is possible because of allele segregation during meiosis.
9. **d** The law of independent assortment states that the alleles for different genes segregate (assort) independently of one another during gamete formation. This is true because those genes are likely on separate chromosomes, which do not influence each other when they separate during meiosis.
10. **False** The events responsible for the laws of segregation and independent assortment occur in meiosis, not mitosis.
11. **c** We would have expected a 3:1 ratio of purple to white in the F_2 generation for this cross instead of the observed 1:1 ratio. However, with a sample size of only 10, significant variation from the predicted ratios is not uncommon. See the Science Toolkit "Tossing Coins and Crossing Plants: Probability and the Design of Experiments" in your textbook for more details about probability and the effect of small sample sizes.
12. **b** Since both parents are homozygotes of opposite genotypes (see Figure 10.9 in the textbook), their offspring must all be heterozygous for the color gene. The presence of a unique phenotype for the heterozygote suggests incomplete dominance as the mechanism of inheritance for this trait.
13. **d** These diseases are caused by multiple genes. They are also strongly influenced by environmental factors such as diet, smoking, and pollution.
14. **e** blending inheritance
    b heterozygote
    d recessive allele
    c incomplete dominance
    a phenotype

Conceptual Understanding

1. **b** There is no evidence that organisms have any control over the kind of mutations that occur in their genetic material.
2. **e** Given that both parents are true-breeding, the cross must be either $AA \times AA$, $AA \times aa$, or $aa \times aa$. If you work out these crosses, you will see that all of the possible answers are true for each possible situation.
3. **c** Heterozygotes can make two kinds of gametes, each with equal probability.
4. **b** Mendelian inheritance still applies in the case of flower color in snapdragons, but this phenomenon is the result of incomplete dominance.
5. **b** The production of coat color in mice is an example of two genes that influence each other, with one having more prominence. There is no environmental control for this trait.
6. **b** Each reproduction event is the result of the fusion of a single gamete from each of the two parents, and therefore they are independent events. Therefore, the probabilities remain the same each time.

7. **e** As with coin tosses, the larger the sample size of fertilizations, the "truer" the result (closer to theoretical predictions) that will be obtained.

8. **d** Since two characters are involved in this cross, both segregation and independent assortment apply.

9. **True** This is a classic example of gene and environment interactions.

10. **False** Since pigment production and deposition in Siamese cats is under the control of a temperature-sensitive enzyme that works better at colder temperatures, just the opposite of this statement would be true.

11. **c** Since both individuals are homozygous recessive for the *c* gene, their offspring will also be homozygous recessive for *c*. Since *c* has more influence over pigmentation than does the *B* gene, their offspring will not deposit coat pigment and will therefore all be white.

12. **b** "True-breeding" in this case means homozygous for both traits, with the parents of opposite phenotypes.

13. **a** The offspring will be heterozygotes for both traits, which means that they will express both dominant phenotypes.

14. **a** In this F_1 genotype, the leaf shape gene is homozygous recessive. This means that both parents would have had to contribute an *r* allele. Since the true-breeding, round-leaved parent does not have such an allele, this F_1 genotype is not possible through the cross described.

CHAPTER 11 Chromosomes and Human Genetics

Getting Started

Below are a few questions to consider prior to reading Chapter 11. These questions will help guide your exploration and assist you in identifying some of the key concepts presented in this chapter.

1. Who was George Huntington, and what was his contribution to the medical field?
2. How many different genes make up the human genome?
3. On which chromosome is the *SRY* gene located, and what role does it play in determining sex in humans?
4. How does the process of crossing-over work to generate nonparental genotypes?
5. Why are genetic disorders that are caused by recessive alleles much more common than those caused by dominant alleles?
6. What are the processes amniocentesis, chorionic villus sampling, and preimplantation genetic diagnosis used for?
7. What do the genetic disorders Down syndrome, Klinefelter syndrome, and Turner syndrome all have in common?

A Guide to the Reading

The following concepts typically give students the most difficulty when exploring the content in Chapter 11 for the first time. For each concept, one or more references have been identified that may help you gain a better understanding of these potentially problematic areas.

Homologous Chromosomes

As discussed in the text, the work of Gregor Mendel and August Weismann suggested that genes were located on physical structures (later determined to be chromosomes) present in the cell. This led to the development of the "chromosome theory of inheritance." The process of inheritance involves the contribution of chromosomes from both parents. Specifically, each individual possesses one complete set of chromosomes from the father and one complete set from the mother. This results in the presence of two copies of each chromosome (and therefore two copies of each gene) in each and every cell. The pairs of chromosomes containing the same genes are referred to as *homologous chromosomes*. During the process of meiosis (Chapter 9), these homologous chromosomes pair at the metaphase plate for proper independent assortment.

For more information on this concept, be sure to focus on:

- Section 11.1, The Role of Chromosomes in Inheritance
- Section 11.4, Origins of Genetic Differences among Individuals

Genetic Linkage and Crossing Over

Despite Gregor Mendel's elaborate work, scientists often observed that certain genes seemed to be inherited together, violating the law of independent assortment. This contradictory observation remained a mystery until the early 1900s when the studies of Thomas Hunt Morgan helped solve the puzzle. Morgan, working with inherited traits in fruit flies, was able to determine that two genes physically located near one another on a chromosome were more likely to be inherited together, violating the law of independent assortment. Of course, this makes sense—we now know that chromosomes are inherited as intact units. Therefore, any genes located on the same chromosome should be inherited together in a linked fashion. However, Morgan also observed that genes that should have been linked would, on occasion, assort independently, producing unexpected phenotypes. This could be explained by the process of crossing-over, where genetic material is exchanged between paired homologous chromosomes during the process of meiosis. The key to understanding crossing-over is to realize that the farther apart two genes are physically located on a chromosome, the more likely it is that crossing-over will occur between these genes. Conversely, the closer two genes are located along a chromosome, the less likely it is that crossing-over will occur between them, resulting in a "tighter" genetic linkage.

For more information on this concept, be sure to focus on:

- Section 11.3, Linkage and Crossing-Over
- Figure 11.4, Some Alleles Do Not Assort Independently
- Figure 11.5, Linkage Is Not Complete

Sex Linkage

Humans possess 23 pairs of homologous chromosomes. Twenty-two of these pairs provide both males and females with 2 copies of each gene. The remaining pair of chromosomes constitute the sex chromosomes (X and Y). Genes located on these chromosomes are considered to be sex-linked since males (XY) would possess only a single copy of each of these genes. As a result, males are particularly susceptible to sex-linked recessive disorders since the inheritance of only a single copy of the defective allele from the mother would result in the presence of the disease. This is in contrast to females, where two copies of the defective allele (one located on each X chromosome contributed by each parent) would be required. Keep in mind that 15 of the 1,200 genes found on the sex chromosomes are, in fact, shared between the X and Y chromosomes. As a result, since 2 copies of these 15 genes are present in both males and females, they would not be considered sex-linked.

For more information on this concept, be sure to focus on:

- Section 11.7, Sex-Linked Inheritance of Single-Gene Mutations
- Figure 11.11, Inheritance of X-linked Recessive Disorders

Tying It All Together

Several concepts presented in this chapter build upon concepts presented in previous chapters and are revisited and discussed in greater detail in subsequent chapters, including:

Noninherited Genetic Disorders

- In Interlude B, An Interplay of Factors Can Cause Cancer

Sickle-Cell Anemia

- Chapter 13 – Section 13.7, Putting It All Together: From Gene to Phenotype

Gene Expression

- Chapter 14 – Section 14.3, Patterns of Gene Expression

DNA Technology

- Chapter 15 – Section 15.3, Applications of DNA Technology
- Chapter 15 – Section 15.4, Ethical Issues and Risks of DNA Technology

Inheritance of Alleles and Evolution

- Chapter 17 – Section 17.4, Mutation: The Source of Genetic Variation
- Chapter 17 – Section 17.7, Natural Selection: The Effects of Advantageous Alleles

Practice Questions

Factual Knowledge

1. Which of the following ideas or events related to the chromosomal theory of inheritance is *false*?
   a. Chromosomes were first observed in dividing cells in 1882.
   b. Mendel was the first to discover that chromosomes contain genes.
   c. Chromosomes are made of a single molecule of DNA and many proteins.
   d. Genes are located in specific places on the chromosome called *loci*.
   e. Homologous chromosomes separate into different gametes during meiosis.

2. Both the Y chromosome in humans and the Z chromosome in birds
   a. are examples of autosomes.
   b. contain the same genes.
   c. are present only in males.
   d. play a role in determining gender.
   e. none of the above

3. In humans, the female parent's genetic contribution determines the sex of a child. (True or False)

4. A person is genetically XX and develops as a male. How can this be explained?
   a. In humans, males are XX.
   b. A mistake of genetic analysis must have taken place, because XX cannot develop into a male.
   c. The *SRY* gene is also present in this individual.
   d. The X chromosomes have nothing to do with sex determination in humans.
   e. none of the above

5. In the absence of crossing over, Mendel's law of independent assortment does not apply when genes are located on the same chromosome. (True or False)

6. On chromosomes, a pair of linked genes will be found only on
   a. autosomes.
   b. sex chromosomes.
   c. the same chromosome.
   d. opposite ends of the same chromosome.
   e. none of the above

7. Genes located on the same chromosome tend to be inherited as a group. Which of the following can disrupt this pattern of inheritance?
   a. sex determination
   b. crossing-over
   c. chromosome replication
   d. changes in chromosome number
   e. all of the above

8. ________ is a process that contributes to genetic variation in organisms.
   a. Crossing-over
   b. Independent assortment
   c. Gene mutation
   d. Fertilization
   e. all of the above

9. Refer to the human pedigree shown in Figure 11.8 of your textbook. The individual labeled "1" in generation III is
   a. a healthy male.
   b. a healthy female.
   c. a male with cystic fibrosis.
   d. a female with cystic fibrosis.
   e. one of the parents.

10. Because humans have long generation times, consciously choose their mates, and produce relatively few offspring, the experimental study of genetic disorders is relatively easy. (True or False)

11. If two parents are heterozygous for an autosomal recessive disease,
    a. they are both considered carriers for the disease.
    b. their children have no chance of inheriting the disease.
    c. their children have a 50 percent chance of inheriting the disease.
    d. all of their children will also be heterozygotes.
    e. all of their children will have the disease.

12. In humans, X-linked genetic diseases
    a. are associated with one of the sex chromosomes.
    b. include hemophilia and Duchenne's muscular dystrophy.
    c. can be seen in both males and females.
    d. tend to be expressed more in males than in females.
    e. all of the above

13. Down syndrome in humans is an example of
    a. normal genetic recombination.
    b. a change in the overall number of chromosomes.
    c. an X-linked recessive disorder.
    d. a single-gene autosomal mutation.
    e. a change in the structure of one of the sex chromosomes.

14. Match each term with the best description.
    __ translocation
    __ crossing-over
    __ locus
    __ pedigree
    __ X-linked
    a. place where a gene is found on a chromosome
    b. movement of a gene to a nonhomologous chromosome
    c. trait associated with one of the sex chromosomes
    d. charts genetic relationships within a family history
    e. exchange of genes between homologous chromosomes

Conceptual Understanding

1. Which of the following statements about homologous chromosomes is *false*?
   a. There are 22 pairs in humans, not counting the sex chromosomes.
   b. They contain the same genes in the same locations.
   c. They pair during meiosis.
   d. They sometimes engage in crossing-over during meiosis.
   e. none of the above

2. Which of the following organisms is capable of producing gametes containing different sex chromosomes?
   a. male butterfly
   b. male human
   c. female human
   d. male bird
   e. none of the above

3. Which of the following blood types would be considered a Universal Donor (capable of donating blood to persons of all blood types)?
   a. A+
   b. O+
   c. AB+
   d. O-
   e. AB-

4. Two genes are known to be on the same chromosome, yet analysis of genetic crosses involving these genes suggests that they assort independently. The most plausible explanation for this observation is that they are
   a. the genes involved in causing Huntington disease.
   b. genes that are especially prone to mutation.
   c. on opposite ends of the same chromosome.
   d. located adjacent to one another on the same chromosome.
   e. X-linked genes.

5. Crossing-over is expected to occur infrequently for genes located very close together on the same chromosome. (True or False)

6. Refer to Figure 11.4 in your textbook. Which result indicates that the genes for body color and wing length in fruit flies are linked?
   a. Four different phenotypes are present in the offspring.
   b. Parental genotypes are present in the offspring.
   c. Nonparental genotypes are present in the offspring.
   d. An unexpectedly low percentage of nonparental genotypes are present in the offspring.
   e. all of the above

7. Which of the following statements regarding crossing-over is *false*?
   a. Crossing-over disrupts the linkage between genes on the same chromosome.
   b. Crossing-over disrupts the linkage between genes on different chromosomes.
   c. Crossing-over produces new genetic combinations.
   d. Crossing-over produces nonparental chromosomes.
   e. none of the above

8. Consider a species that reproduces sexually and has a total of 3 pairs of chromosomes (2 pairs of autosomes and 1 pair of sex chromosomes). In the absence of crossing-over, how many different offspring combinations can a mating pair produce?
   a. 4
   b. 8
   c. 9
   d. 16
   e. 64

9. Consider the pedigree for a family with cystic fibrosis shown in Figure 11.8 of your textbook. On what basis can we conclude that the allele that causes this disease is recessive?
   a. No one in the entire pedigree has the disease.
   b. No one in generation I has the disease.
   c. Neither individual in generation II has the disease.
   d. Only one of the males in generation III has the disease.
   e. Two-thirds of those with the disease in generation III are females.

10. Sickle-cell anemia is an inherited chronic blood disease caused by an autosomal recessive allele. Suppose a man who is homozygous recessive for the sickle-cell gene fathers a child by a woman who is a carrier for sickle-cell. What are the chances their children will exhibit the disease? (Note: A Punnett square may be useful.)
    a. 0%
    b. 25%
    c. 50%
    d. 75%
    e. 100%

11. Autosomal dominant diseases are exhibited by anyone who carries at least one dominant allele for that gene. How is it that dominant lethal genes, such as the one that causes Huntington disease, can persist in a population?
    a. The disease-causing allele can "hide" in the heterozygous condition.
    b. The disease develops only under the influence of other genes.
    c. The environment plays a large role in determining whether or not the gene is expressed.
    d. These diseases usually take effect later in life after people have had children.
    e. all of the above

12. Consider the Congenital Generalized Hypertrichosis (CGH) disorder (Figure 11.12). What can you determine about the potential offspring of a couple in which the father has CGH and the mother does not?
    a. All sons will have the disorder.
    b. One-half of the sons will have the disorder.
    c. All daughters will have the disorder.
    d. One-half of the daughters will have the disorder.
    e. We cannot determine the outcome without knowing the genotype of the mother.

13. Cri du chat syndrome (Fig. 11.14) is caused by which of the following structural changes to chromosomes?
    a. deletion
    b. duplication
    c. inversion
    d. translocation
    e. trisomy

NOTE: For more practice using the concepts of human genetics, we strongly encourage you to work through the sample genetics problems at the end of Chapter 11 in your textbook.

Related Activities

- Imagine that you are in the role of having to counsel a teenager who has just learned that Huntington disease runs in her family. Describe how you would explain the biology of the disease to this person, including the genetics determining her chances of acquiring the disease and chances of passing on the disease. Would you encourage her to undergo genetic screening to find out if she has the defective gene? Why or why not?

- Using the World Wide Web or resources from your library, research the basis of an X-linked recessive disorder in humans. Possibilities include color-blindness, hemophilia, and male pattern baldness, but there are many others. Compose a one-page summary that explains how the disorder manifests itself, and provide a hypothetical pedigree to chart the inheritance of the disorder through at least three generations of a family.
- Consider the discussion of the genetics of myxomas presented in the Biology in the News section at the end of the chapter. In this instance, individuals that are afflicted with a rare muscle disorder also tend to develop these potentially dangerous heart tumors. Given what you have learned from this chapter regarding genetic linkage, compose a short one-paragraph explanation of how two seemingly disparate disorders may be inherited together.

Answers and Explanations

Factual Knowledge

1. **b** The chromosomal theory of inheritance, including the discovery that genes are on chromosomes, was developed long after Mendel had died.
2. **d** Recall from section 11.2 that W, X, Y, and Z are all types of sex chromosomes. Human males are XY, and male birds are ZZ. The W chromosome is present only in female birds.
3. **False** The human male produces two different kinds of gametes with respect to the sex chromosomes (X or Y), whereas the female produces one (X only). It is therefore the male who determines the gender of offspring in our species.
4. **c** It is possible (although rare) for the portion of a Y chromosome that determines maleness (the *SRY* gene) to become associated with an X chromosome, and thus be transmitted from father to daughter.
5. **True** Linked genes (those present on the same chromosome) do not show independent assortment, unless they are located very far apart from one another on the chromosome so that crossing-over is likely to occur.
6. **c** Genetically linked genes are present on the same chromosome, regardless of type.
7. **b** Recall that the process of crossing-over can switch alleles between homologous chromosomes.
8. **e** All of these processes would contribute to the generation of genetic differences between individuals.
9. **b** Refer to the key that accompanies Figure 11.8.
10. **False** Because of these factors, as well as matters of ethical concern, it is impossible to do controlled "breeding" of humans for purposes of genetic analysis. Pedigrees are the best way to study inheritance in our species.
11. **a** Heterozygotes have one dominant (in this case, normal) and one recessive (in this case, disease-causing) allele. Therefore, they do not show the disease, but can pass along the disease-causing allele to their offspring. If two heterozygotes mate, on average one-fourth of their children will have the homozygous recessive genotype and consequently be affected by the disease.
12. **e** Since X chromosomes are present in males and females, both can show X-linked diseases. Most X-linked genetic diseases are recessive in nature. Males express them more frequently, however, since they only need one copy of the defective gene, instead of two for females.
13. **b** Down syndrome results when an individual receives three copies of chromosome 21 duc to an error during meiosis.
14. **b** translocation
    e crossing-over
    a locus
    d pedigree
    c X-linked

Conceptual Understanding

1. **e** All of the statements are true regarding homologous chromosomes.
2. **b** Recall that human males are XY with respect to the sex chromosomes, while females are XX. In birds and butterflies, males are ZZ.
3. **d** Recall from the Biology Matters section of the chapter and Chapter 10 that blood types are determined according to the presence of marker genes on the surface of red blood cells. Individuals with the recessive type O^- blood do not express any of these markers, and therefore are capable of donating blood to everyone.
4. **c** Even though genes located on the same chromosome are technically linked, genes present at opposite ends of the same chromosome would be so far apart that they have a high probability of crossing over and therefore behaving as separate chromosomes.

5. **True** The closer together two genes are, the less likely they are to be involved in a crossing-over event.

6. **d** If these genes were not linked, we would expect a much higher percentage of nonparental genotypes than those observed.

7. **b** Recall that crossing-over occurs only between homologous chromosomes and that linked genes are found only on the same physical chromosome. Therefore, these processes do not impact genes found on different chromosomes.

8. **e** Each parent would be able to produce gametes with $2^3 = 8$ different combinations of chromosomes. In a mating event, there would therefore be $8 \times 8 = 64$ possible combinations of offspring.

9. **c** If the disease allele were dominant, it would be expressed in at least one of the two parents of the afflicted individuals. Since both parents in Generation II are normal, this indicates they are carriers of the recessive allele. The reason Generation I does not inform us is the possibility that the mutated allele might have first arisen in generation II.

10. **c** The cross is *Ss* × *ss*, which yields the same two genotypes in the offspring. Half will be carriers; the other half will have sickle-cell disease.

11. **d** Late onset of the disease, after breeding has occurred, explains why people pass along this otherwise lethal mutation.

12. **c** CGH is an X-linked, dominant trait. Since the father has CGH, this means he must have the affected X-chromosome. Since all of his daughters will receive his X-chromosome, they will be affected, while none of his sons will receive his X-chromosome, and therefore will be unaffected.

13. **a** Cri du chat results from the loss of a critical portion of chromosome number 5. This type of event is referred to as a *deletion* (see Figure 11.13).

CHAPTER **12** DNA

Getting Started

Below are a few questions to consider prior to reading Chapter 12. These questions will help guide your exploration and assist you in identifying some of the key concepts presented in this chapter.

1. In the early 1900s, what molecule did most geneticists believe genes were composed of?
2. How did a pneumonia-causing bacterium contribute to our understanding of the molecular basis of the genetic material?
3. In terms of genetics, what is transformation?
4. Who was Rosalind Franklin, and what was her contribution to the study of DNA?
5. In terms of the DNA molecule, what is a complementary strand?
6. How is the process of DNA replication similar to word processing on a computer?
7. How many rads of radiation energy are required to mortally injure a human?
8. Why must people with xeroderma pigmentosum stay out of the sun?
9. By what age does a typical person accumulate more than half of his or her cancer risk as a result of exposure to cancer-causing agents?

A Guide to the Reading

The following concepts typically give students the most difficulty when exploring the content in Chapter 12 for the first time. For each concept, one or more references have been identified that may help you gain a better understanding of these potentially problematic areas.

Search for the Genetic Material

The experiments of Frederick Griffith described in the textbook (Figure 12.1) highlight the transformation of the pneumonia-causing bacterium *Streptococcus pneumoniae*. The key to understanding these experiments is to realize how the two different strains used by Griffith affect mice when injected. The "S" strain (smooth) is more deadly than the "R" strain (rough) and can also be rendered harmless through heat-treatment. The critical experiment conducted by Griffith involved mixing living bacteria of the harmless "R" type with the equally harmless heat-killed "S" strain. When the mixture was injected into mice, they unexpectedly died. Based on this result, Griffith proposed that some factor remaining from the heat-killed "S" strain was capable of transforming the living, harmless "R" strain bacteria into a virulent type. This transforming factor was believed to be the genetic material. The exact chemical nature of this factor remained a mystery until the experiments conducted by Hershey and Chase (Figure 12.2) using radioactively labeled virus particles. Since it was possible for Hershey and Chase to selectively label protein and DNA separately, this allowed them to trace which of these two molecules was responsible for transforming infected bacterial cells into tiny virus factories. As seen in the figure, it was determined that DNA was in fact the genetic material.

For more information on this concept, be sure to focus on:

- In Section 12.1, Harmless bacteria can be transformed into deadly bacteria
- In Section 12.1, The genetic instructions of a virus are contained in its DNA
- Figure 12.1, Genetic Transformation of Bacteria
- Figure 12.2, DNA Is the Genetic Material

DNA Replication

The key to deciphering how DNA replication occurs came with the discovery of the 3-dimensional structure of the DNA molecule. Recall that the DNA molecule consists of a pair of DNA strands, each strand complementary to the other. The two strands are held together via hydrogen-bonding between the complementary bases. These hydrogen bonds form a base pair. In order for replication to proceed, the two strands must be separated, breaking the hydrogen bonds in the process. Once separate, each of the two strands then contains the required information to serve as a template for the construction of a new complementary DNA strand.

For more information on this concept, be sure to focus on:

- In Section 12.2, DNA is a double helix
- Section 12.3, How DNA Is Replicated
- Figure 12.5, DNA Replication

DNA Repair

DNA repair is vital to the survival of all organisms. The need for an efficient DNA repair mechanism is due to the many varied ways in which DNA may become damaged. As discussed in the chapter, this can occur when simple mistakes are made during the process of DNA replication or as the result of physical damage caused by radiation or chemicals. In all cases, whenever damage occurs, the cell must recognize the damage, act to remove the damaged segment of DNA, and work to replace it with the correct information. This involves the action of several specialized enzymes, all working together to maintain the accuracy of the genetic information.

For more information on this concept, be sure to focus on:

- In Section 12.4, Few mistakes are made in DNA replication
- In Section 12.4, Normal gene function depends on DNA repair
- Figure 12.7, Repair Proteins Fix DNA Damage

Tying It All Together

Several concepts presented in this chapter build upon concepts presented in previous chapters and may also be revisited and discussed in greater detail in subsequent chapters, including:

Nucleotides

- Chapter 4 - in Section 4.5, Nucleotides store information and energy

Cancer

- In Interlude B, An Interplay of Factors Can Cause Cancer

Inherited Genetic Disorders

- Chapter 11 - Section 11.5, Human Genetic Disorders

The Genetic Code

- Chapter 13 - Section 13.2, How Genes Control the Production of Proteins
- Chapter 13 - Section 13.4, The Genetic Code

Gene Expression

- Chapter 14 - Section 14.4, How Cells Control Gene Expression

Practice Questions

Factual Knowledge

1. During the first half of the twentieth century, as geneticists began the search for what genes are made of, they knew that the hereditary material
   a. was contained in cells and was passed from one generation to the next.
   b. must contain all of the information necessary for life.
   c. had to be composed of a material that could be accurately replicated.
   d. needed to be sufficiently variable to accommodate all life forms.
   e. all of the above
2. From the very start, DNA was thought to be a good candidate for the hereditary material because geneticists believed it was a complex molecule. (True or False)
3. Which line of research did *not* provide some evidence that DNA is the hereditary material?
   a. studies of how certain viruses attack their bacterial hosts
   b. exposing one strain of bacteria to isolated DNA from other strains
   c. using heat-killed bacteria to transform other strains of living bacteria
   d. radioactively labeling viral DNA and protein
   e. none of the above
4. Many researchers contributed information that ultimately led to the discovery of the molecular structure of DNA. Which of the following was *not* one of these researchers?
   a. Francis Crick
   b. Linus Pauling
   c. Rosalind Franklin
   d. James Watson
   e. Maurice Wilkins
5. The chemical bonds that hold the two complementary strands of the DNA double-helix close to one another are covalent bonds. (True or False)
6. The primary purpose of Griffith's experiments on the *Streptococcus pneumoniae* bacterium was to
   a. find a cure for pneumonia in humans.
   b. prevent cancers caused by exposure to ultraviolet light.

c. determine if DNA is the hereditary material.
d. discover the molecular structure of DNA.
e. determine which DNA nucleotides bond with one another.

7. The key finding of the Hershey and Chase experiments on the mechanism of viral replication was that
   a. protein, not DNA, is the hereditary material.
   b. DNA, not protein, is the hereditary material.
   c. protein and DNA play an equal role in determining inheritance.
   d. neither protein nor DNA plays a role in determining inheritance.
   e. both sulfur and phosphorous must be present for viral DNA to replicate.

8. Which of the following parts of a DNA molecule are held together by hydrogen bonds?
   a. the carbons within the sugar–phosphate group
   b. the carbons within the nitrogen-containing bases
   c. nucleotide bases on opposite strands of the helix
   d. successive nucleotides within a single strand of the helix
   e. all of the above

9. Consider the three-dimensional structure of DNA. Which of the following has a structure analogous to a molecule of DNA?
   a. railroad tracks
   b. painted lines on a highway
   c. spiral staircase
   d. concrete sidewalk
   e. curly drinking straw

10. Even though you and your brothers or sisters are all humans, odds are that each of you has a somewhat different genetic makeup. The molecular basis for this inherited genetic variation
    a. is related to the difference in the DNA nucleotide sequence of your genes.
    b. depends on whether protein or DNA is used as the hereditary material.
    c. depends on whether adenine pairs with thymine or guanine.
    d. is the result of errors in the DNA replication process.
    e. has nothing to do with DNA.

11. When a molecule of DNA replicates without error, each of the resulting molecules contains
    a. the same amount of A as T.
    b. the same amount of G as C.
    c. one new strand and one old strand.
    d. the same sequence of bases as did the original molecule.
    e. all of the above

12. Which of the following statements about the process of DNA replication is *false*?
    a. Many different enzymes are needed for the process to function properly.
    b. Mistakes can be corrected at multiple steps in the process.
    c. Uncorrected mistakes introduce mutations into the DNA base sequence.
    d. Mistakes in the copying process are very common occurrences.
    e. Mutations represent a source of new alleles.

13. Match each term with the best description.
    __ DNA bases
    __ DNA replication
    __ mutation
    __ nucleotide
    __ transformation
    a. specific change in DNA base sequence
    b. adenine, thymine, cytosine, guanine
    c. sugar–phosphate group bonded to a nitrogen base
    d. mechanism for copying genes
    e. cell genotype changes after exposure to DNA from another organism

Conceptual Understanding

1. The primary reason DNA was first thought to be a poor candidate for the hereditary material was that
   a. Griffith's experiments showed that protein, not DNA, caused transformation.
   b. viruses lack DNA yet still pass genetic information between generations.
   c. DNA was believed to have a simple chemical structure with little variability.
   d. the work of Hershey and Chase showed that protein was the genetic material.
   e. none of the above

2. Consider Griffith's experiments on transformation in *Streptococcus pneumoniae*. Now imagine you are extending these experiments by injecting a mixture of heat-killed strain R bacteria and live strain S bacteria into a mouse. You would predict that the mouse would ________, and you will find live strain ________ bacteria in its blood.
   a. die; R
   b. live; R
   c. die; S
   d. live; S
   e. none of the above

3. Which of the following is the most important result from the experiments of Avery and colleagues?
   a. There are multiple strains of pneumonia-causing bacteria.
   b. Only DNA is able to transform bacteria from one type to another.
   c. Only protein is able to transform bacteria from one type to another.
   d. The hereditary material of viruses is DNA.
   e. DNA is a double-stranded molecule twisted into a helix.

4. In the Hershey and Chase experiments on the mechanism of viral reproduction, only radioactive phosphorus associated with the viral DNA ended up inside the host bacterium. (True or False)

5. Assume a certain molecule of DNA is composed of exactly 22 percent adenine. How much cytosine would you expect to find in this molecule?
   a. 22%
   b. 44%
   c. 56%
   d. 28%
   e. 24%

6. A short sequence of bases on one strand of DNA is AGTCTACCGATAGT. If this sequence serves as a template for the formation of a new strand of DNA, what will be the corresponding base sequence in the new strand?
   a. AGTCTACCGATAGT
   b. TCAGATGGCTATCA
   c. TGATAGCCATCTGA
   d. GACATCGATTCGAT
   e. THISISNOTCORRECT

7. A certain enzyme is known to be responsible for separating the two strands of a DNA molecule to facilitate the process of DNA replication. This enzyme most likely
   a. breaks covalent bonds within nucleotides.
   b. causes the DNA to twist into a helix.
   c. forms covalent bonds between nucleotides.
   d. adds a sugar–phosphate group to nucleotides.
   e. breaks hydrogen bonds between bases.

8. In 1951, Linus Pauling became the first person to determine the three-dimensional structure of a protein. His work suggested that
   a. it should also be possible to determine the three-dimensional structure of DNA.
   b. proteins are the hereditary material.
   c. DNA is the hereditary material.
   d. DNA is a chemical derivative of protein.
   e. none of the above

9. In an earlier chapter you learned that DNA analysis is one of the best ways to determine the evolutionary relationships among organisms. Which of the following pairs of organisms should therefore have the most similar sequence of bases in their DNA?
   a. chicken and human
   b. dragonfly and bacterium
   c. fish and birds
   d. chimpanzee and human
   e. virus and mushroom

10. Assume that you chemically label both strands within a molecule of DNA. You then allow this DNA to replicate using unlabeled nucleotides. Which of the following statements about the two resulting DNA molecules would be *false*?
    a. Both molecules would contain the chemical label.
    b. One molecule would contain the chemical label, the other would not.
    c. One strand within each molecule would contain the chemical label.
    d. Assuming no replication errors, both molecules would be genetically identical.
    e. none of the above.

11. An uncorrected mismatch error during human DNA replication can result in
    a. the potential for a disease or genetic disorder.
    b. the combination of two or more alleles.
    c. a disruption in the double-helix structure of the DNA molecule.
    d. the introduction of many new mismatch errors.
    e. all of the above

12. Before damaged DNA can be repaired, which of the following must occur?
    a. The damaged area must be recognized.
    b. The damaged area must be removed.
    c. The damaged area must be replaced.
    d. A collection of various repair enzymes must be present.
    e. all of the above

13. The genetic disorder xeroderma pigmentosum (XP) is a tragic example of the failure of DNA repair mechanisms. Specifically, XP patients possess a ________ allele that produces a nonfunctional protein whose job it would normally be to repair DNA damage caused by ________.
    a. dominant; ultraviolet light
    b. dominant; arsenic
    c. recessive; ultraviolet light
    d. recessive; PAHs
    e. dominant; cancer

14. Assume a cell present in your body has performed a mismatch error during the process of DNA replication. This error has resulted in the generation of a new allele for one of your genes. In order for this new allele to be inherited by your offspring, which of the following conditions must be *true*?
   a. The error must be repaired.
   b. The error must be caused by a chemical mutagen.
   c. The error must occur in the germ line cells responsible for the production of gametes.
   d. The new allele produced must be dominant.
   e. The new allele produced must be recessive.

Related Activities

- Although you may conceptually understand the process of DNA replication, there is no substitute for actually working with physical models and demonstrating mechanically how new DNA can be built from an existing template. Using paper shapes, create twenty models of each of the four DNA bases. (The exact shape doesn't matter, as long as the four bases are different from one another. Making photocopies of the shapes will save you some work!) Construct a molecule of DNA that is ten base pairs long. Figure out how this molecule can serve as a template for the formation of two identical molecules of DNA. (Hint: You might use the DNA segment shown in Figure 12.4 as a starting point.) Once you have mastered basic DNA replication, practice introducing mutations to see how these changes affect the replication of new DNA molecules.
- The Biology Matters box ("Cancer in Children") in this chapter highlights the risk of exposure to carcinogens (environmental mutagens that cause cancer) and the development of cancer in young children. Search the World Wide Web to locate a specific example of environmentally induced cancer in children, and compose a one-page essay about the relative risk endured by children as compared to adults for this example. Be sure to include what is known about how the carcinogen actually interacts with DNA and ultimately leads to cancer.
- Until very recently, the history of science was full of examples of intelligent and capable women who were either prevented from becoming scientists or, if they did manage to do research, had their work ignored simply because they were female. Rosalind Franklin, discussed in this chapter, is only one such example of this unfortunate gender bias in science. Consult an encyclopedia or other reference source and write a one-page essay about another woman scientist who struggled to have her work properly recognized in the male-dominated world of research.

Answers and Explanations

Factual Knowledge

1. **e** The work of Mendel and others had already provided this information and set the stage for the molecular study of genetic inheritance that ultimately led to the discovery of the structure of DNA.
2. **False** Prior to the work of Franklin, Wilkins, Watson, and Crick, the structure of DNA was incorrectly thought to be relatively simple. Geneticists therefore did not believe it could provide the necessary complexity to be the hereditary material. Because of its chemical diversity, protein was initially believed to be the best candidate.
3. **c** Although bacterial transformation is in fact due to DNA, Griffith's initial experiments demonstrating this phenomenon did not provide evidence that DNA itself was responsible for transformation since both protein and DNA were present in the extracts.
4. **b** Linus Pauling was the first to determine the three-dimensional structure of protein, not DNA.
5. **False** While each separate DNA strand is held together by strong covalent bonds, the two complementary strands are kept in close proximity to one another with relatively weak hydrogen bonds.
6. **a** Griffith was a physician trying to find a way to effectively treat or cure pneumonia. Nothing much was known about the structure or function of DNA at the time he did his work.
7. **b** Hershey and Chase used radioactive labeling of sulfur and phosphorus to clearly demonstrate that DNA, and not protein, is the material that carries hereditary information. Review Figure 12.2 for additional information.
8. **c** Hydrogen bonds connect the bases on opposite strands of the DNA molecule, which is the reason that the helix can be unwound and separated relatively easily. All other bonds in the molecule are covalent.
9. **c** The DNA helix is structurally very similar to a spiral staircase, with the hydrogen bonds between bases representing the steps and the railings corresponding to the opposing strands of nucleotides.
10. **a** The fundamental factor that makes one individual different from another is the specific sequence of DNA bases in our genes.

11. **e** All of these will be true if the DNA molecule replicates correctly.

12. **d** Even though DNA replication typically occurs millions of times during the life of a multicellular organism, it is a remarkably accurate process. Those errors that do occur are usually corrected with a high degree of reliability.

13. **b** DNA bases
    d DNA replication
    a mutation
    c nucleotide
    e transformation

Conceptual Understanding

1. **c** Since the building blocks of DNA were thought to be structurally simple and few in number (4), biologists prior to the 1940s thought that only proteins could have the necessary complexity and diversity to be the hereditary material.

2. **c** Since you are directly injecting live S strain into the mouse and this strain is virulent, you would expect that the mouse would die, and you would find living strain S bacteria in its blood. The heat-killed R strain would have no impact on this experiment.

3. **b** Avery and coworkers isolated pure DNA and showed that only it could cause bacterial transformation. Their work did not address the actual structure of DNA.

4. **True** Since phosphorus is abundantly present in DNA, the fact that radioactive phosphorus from the virus was present inside the bacterium meant that DNA must have been injected into the host cell. Recall that these researchers also selectively tagged viral protein using radioactive sulfur, which is present in protein, but not DNA.

5. **d** Recall that since A pairs with T, there should be equal amounts of both bases in the molecule, totaling 44%. Therefore, the remaining 56% must be made from equal amounts of C and G, meaning we would expect half this amount (28%) to be the total percentage of cytosine present in the molecule.

6. **b** Following the Watson-Crick base pairing rules (A-T; G-C), we can easily determine the sequence of the complementary strand.

7. **e** Hydrogen bonds between complementary bases on opposite strands hold the DNA molecule together. An enzyme that separates the strands must therefore break these bonds.

8. **a** Pauling's work on protein structure provided the molecular groundwork that paved the way for determining the three-dimensional structure of other complex biological molecules such as DNA.

9. **d** Being members of species in closely related groups, chimpanzees and humans would share the most similar DNA sequences. In fact, our DNA is over 95% similar when compared to that of a chimpanzee.

10. **b** In DNA replication, each strand of the original molecule serves as a template for the formation of a new strand. Thus, both new molecules would have half of its nucleotides (and the chemical label) from the original molecule.

11. **a** The introduction of a mismatch error has the potential of changing a particular allele permanently. In some instances, this change can result in a defective gene that may then serve as the basis of a genetic disorder.

12. **e** DNA repair is a multi-step process involving all of the events and structures listed.

13. **c** In XP, a recessive allele codes for the production of a faulty protein in the normal DNA repair mechanism. Thus, when UV light damages DNA in skin cells, the damage cannot be repaired and the physical symptoms of XP, such as severe blisters and skin cancer, emerge.

14. **c** Keep in mind that in order for an allele to be passed on to the next generation, it must be present in either the sperm or the egg cells contributed by the parent. If the mutation occurs elsewhere in the body, it will have no chance of being inherited by subsequent generations.

CHAPTER 13 From Gene to Protein

Getting Started

Below are a few questions to consider prior to reading Chapter 13. These questions will help guide your exploration and assist you in identifying some of the key concepts presented in this chapter.

1. What did British physician Archibald Garrod believe was the cause of inherited human metabolic disorders such as alkaptonuria?
2. What are the functional differences between the three types of RNA—mRNA, rRNA, and tRNA?
3. What is a codon?
4. What is the primary difference between an intron and an exon?
5. In the genetic code, what sequence constitutes the universal start codon?
6. How are frameshift mutations similar to losing your place on a multiple-choice answer sheet?
7. What is a silent mutation?
8. Which genetic disease results in the formation of curved and distorted red blood cells?
9. What are the dangers associated with "GTC" repeats?

A Guide to the Reading

The following concepts typically give students the most difficulty when exploring the content in Chapter 13 for the first time. For each concept, one or more references have been identified that may help you gain a better understanding of these potentially problematic areas.

Three Types of RNA

The cell makes use of three different types of RNA (messenger, transfer, and ribosomal), each with a specific function associated with the production of proteins from genes. Messenger RNA represents a copy of the sequence of nucleotides present in a gene. This information is then decoded to produce proteins by structures called *ribosomes*. A major component of ribosomes is ribosomal RNA. The building blocks of proteins are amino acids. Amino acids are transferred to the ribosome on transfer RNA molecules for use in protein synthesis. Each of the three types of RNA molecules has a very specific role in the process of gene expression.

For more information on this concept, be sure to focus on:

- In Section 13.1, Three types of RNA are involved in the production of proteins
- Table 13.1, RNA Molecules and Their Functions

Transcription

The first step in the process of gene expression is the "copying" of the genetic information stored in the DNA molecule into a form that may be used in the cytoplasm. This copy is made using RNA, resulting in the production of a messenger RNA (mRNA) molecule. It is important to note that only one of the two complementary strands of the DNA molecule contains the proper sequence used for the production of the mRNA molecule. The strand that contains the information is referred to as the *template strand*. The starting point for a gene is marked by a specific sequence called the *promoter*, so named because it "promotes" the attachment of the RNA polymerase molecule responsible for producing the mRNA molecule. It is also important to remember that during transcription, the ribonucleotide uracil is substituted for the deoxynucleotide thymine. RNA polymerase continues to produce the mRNA molecule, following the sequence present in the gene, until it reaches the "terminator"—a specific sequence that signals the completion of the transcription process. This process results in a mRNA molecule composed of both *introns* and *exons*. The introns contain extraneous information and must be removed prior to translation, completing the formation of a complete mRNA molecule.

For more information on this concept, be sure to focus on:

- Section 13.3, Transcription: Information Flow from DNA to RNA
- Figure 13.3, An Overview of Transcription
- Figure 13.4, Removal of Introns by Eukaryotic Cells

Translation

The second step in the process of gene expression is the decoding of the genetic information copied from the genes into its final form, protein. Information decoding is performed by specialized structures present in the cytoplasm called *ribosomes*. Ribosomes attach to the mRNA molecule (produced through transcription) and begin the process of translating the sequence of nucleotides using the genetic code (Figure 13.6) into a sequence of amino acids resulting in the formation of an intact protein. In addition to mRNA, ribosomes must also interact with tRNA molecules that transport amino acids. Transfer RNA molecules provide the key to translating the genetic code. These molecules contain a three-nucleotide sequence (called the anticodon) that is complementary to the codon sequences present in the mRNA molecule. The codon and anticodon match precisely using Watson-Crick base pairing rules (Chapter 12). Since there are 64 possible combinations of codons, this means there are a total of 64 different tRNA molecules, each with a complementary anticodon. Each of these different tRNA molecules carries only one of the 20 different amino acids. Therefore, it is the process of tRNA production that matches anticodon sequences to the appropriate amino acid molecule, ensuring the accuracy of the translation process.

For more information on this concept, be sure to focus on:

- Section 13.4, The Genetic Code
- Section 13.5, Translation: Information Flow from mRNA to Protein

- Figure 13.6, The Genetic Code
- Figure 13.7, Transfer RNA
- Figure 13.8, Translation

Mutations

As discussed in the current and previous chapters, mutations alter a gene sequence, which may result in the formation of a new allele. The chapter mentions the three main types of mutations that may occur. Substitution mutations are caused by the substitution of one nucleotide with another (such as the substitution of a T for a C). This may occur as an error made during DNA replication. Not all substitution mutations are significant in terms of gene expression. In order for such a mutation to affect a gene product, the error must occur not only within a gene, but within a coding region (exon) of the gene. If the error occurs within an intron region, this segment will be removed prior to translation, and therefore will have no impact on the protein product. In some cases, a substitution mutation may also be "silent." Silent mutations are possible because of the redundancy of the genetic code (more than one codon may specify the same amino acid). For example, the codons GAA and GAG both code for the amino acid Glutamate. As a result, if a substitution mutation were to occur in the third position (A for G or G for A), the amino acid would remain unchanged in the final protein product. Insertion and deletion mutations are potentially more disruptive. In these cases, nucleotides are inserted or deleted into the gene sequence. This may result in a genetic "frameshift" where the codons become shifted by one base, disrupting the meaning of the information. For example, consider the two codon sequence GAA-GAG. In the case of an insertion mutation at the second position (inserting a G), the sequence would then become GGA-AGA-G. If you were to translate the two sequences using the genetic code, you will see that the codons now specify quite different amino acids.

For more information on this concept, be sure to focus on:

- Section 13.4, The Genetic Code
- In Section 13.6, Mutations can alter one or many bases in a gene's DNA sequence
- In Section 13.6, Mutations can cause a change in protein function
- Figure 13.6, The Genetic Code
- Figure 13.9, Effects of DNA Mutations on Protein Production

Tying It All Together

Several concepts presented in this chapter build upon concepts presented in previous chapters and may also be revisited and discussed in greater detail in subsequent chapters, including:

Protein Synthesis

- Chapter 4 – in Section 4.5, Amino acids are the building blocks of proteins

Gene Mutation

- Chapter 10 – Section 10.2, Gene Mutations: The Source of New Alleles

Chromosomal Basis of Inheritance

- Chapter 11 – Section 11.1, The Role of Chromosomes in Inheritance

Genetic Disorders

- Chapter 11 – Section 11.5, Human Genetic Disorders

DNA Replication

- Chapter 12 – Section 12.3, How DNA Is Replicated

Control of Gene Expression

- Chapter 14 – Section 14.4, How Cells Control Gene Expression

Practice Questions

Factual Knowledge

1. In order for the information contained in a gene to be used to produce a functioning protein
   a. DNA must be replicated.
   b. information must be transcribed into mRNA and then translated into amino acids.
   c. tRNA must be transcribed into rRNA and then translated into amino acids.
   d. ribosomes must be converted from rRNA into mRNA.
   e. the mRNA message must be sent from the cytoplasm to the cell nucleus.

2. Genes code for which of the following biological molecules?
   a. enzymes
   b. structural proteins
   c. transport proteins
   d. defense proteins
   e. all of the above

3. A sequence of DNA that contains information for the synthesis of RNA molecules used in the manufacture of proteins is also known as a(n) _______.
   a. intron
   b. mutation
   c. gene
   d. polymerase
   e. codon

4. The process of gene transcription begins with the
   a. binding of RNA polymerase to a region of DNA called the promoter.
   b. removal of introns from the newly formed mRNA.
   c. joining of rRNA with various ribosomal proteins.
   d. attachment of an mRNA molecule to the ribosome.
   e. insertion of new DNA bases into the gene.

5. Which of the following events does *not* occur during the process of translation?
   a. The information in mRNA is used to guide the assembly of amino acids into proteins.
   b. (C)ytosine in DNA pairs with (G)uanine in RNA, and (T)hymine in DNA pairs with (A)denine in RNA.
   c. Start and stop codons present in the mRNA molecule help control the process.
   d. Amino acids are attached to the end of the new protein molecule under construction.
   e. The ribosome interacts with mRNA in the cytoplasm.

6. The coding strand in DNA
   a. is identical in sequence to the template strand.
   b. is translated by ribosomes into functional proteins.
   c. is duplicated in the form of mRNA during transcription.
   d. contains only exons.
   e. none of the above

7. There are two binding sites on a molecule of tRNA, one for the attachment of an amino acid, and the other where the tRNA anticodon binds to the corresponding codon on the mRNA molecule. (True or False)

8. Refer to Figure 13.8 in your textbook. The object shaped like a triangle represents
   a. the last amino acid added to the new protein.
   b. the first amino acid added to the new protein.
   c. the anticodon of the last tRNA used to attach a new amino acid.
   d. RNA polymerase.
   e. a codon.

9. In a ribosome in the process of translating a molecule of mRNA, a maximum of _______ codon(s) can be occupied by tRNA at any one time.
   a. 1
   b. 2
   c. 3
   d. 64
   e. depends on the kind of mRNA

10. Consider the DNA template shown in the middle of Figure 13.9 in your textbook. If the C at the start of the sequence of bases is deleted, this genetic mutation will result in
    a. a frameshift that affects the entire mRNA molecule.
    b. a change in the first amino acid coded for, but not in any others.
    c. a change in the last amino acid coded for, but not in any others.
    d. all the amino acids coded for will now be the same.
    e. no effect on either the mRNA or the protein manufactured.

11. Similar to genes that code for proteins, the products of genes that code for rRNA and tRNA must also undergo translation. (True or False)

12. Match each term with the best description.
    __ codon
    __ mRNA
    __ transcription
    __ translation
    __ intron
    a. conversion of mRNA information to amino acid sequence
    b. 3-base combination of nucleotides in mRNA
    c. occurs within the cell nucleus
    d. unused portion of an mRNA transcript
    e. specifies sequence of amino acids in proteins

Conceptual Understanding

1. A murder has occurred, and you are asked to help solve it. The police bring you a sample from the crime scene of what they believe is the killer's DNA and ask you for a chemical analysis. Your study of this sample reveals the presence of adenine, thymine, ribose, and uracil, leading you to conclude that the sample is
    a. pure DNA.
    b. pure RNA.
    c. probably a mixture of DNA and RNA.
    d. probably a mixture of rRNA and mRNA.
    e. definitely not from the killer.

2. The work of Garrod and Bateson on inherited human metabolic disorders provided some of the first clues that genes work by controlling the production of proteins. (True or False)

3. If you expose a cell to chemicals that specifically disrupt the function of RNA polymerase, which of the following processes will be most directly impacted?
    a. transcription
    b. translation
    c. DNA replication
    d. rate of mutation
    e. evolution of the genetic code

4. Refer to the information on the genetic code provided in Figure 13.6. Use this information to determine how many amino acids are coded for by the mRNA sequence AAUCGCAGUCGGUAG.
    a. 0
    b. 4
    c. 5
    d. 15
    e. cannot determine from the information given

5. A small segment of DNA on the template strand contains the base sequence CGT. If an mRNA transcript is made that includes this sequence, what would be the anticodon on the tRNA that would bind to this corresponding mRNA sequence?
    a. CGT
    b. GCA
    c. CGU
    d. GCT
    e. cannot determine from the information given

6. Without the ________ site on a molecule of DNA, ________ cannot bind to the DNA to begin the process of ________.
    a. active; introns; tRNA translation
    b. promoter; RNA polymerase; mRNA transcription
    c. promoter; rRNA; ribosome activation
    d. attachment; ribosomal proteins; frameshift mutation
    e. stop codon; amino acids; translation

7. Consider a complete, functional piece of mRNA that contains a total of 66 codons. What is the maximum number of amino acids that could be present in the protein coded for by this mRNA?
    a. 21
    b. 22
    c. 64
    d. 65
    e. 66

8. A triplet base sequence on the template strand of DNA reads ATT. What will be the corresponding mRNA codon, tRNA anticodon, and amino acid called for by this DNA?
    a. TAA; UTT; methionine
    b. TAA; AUU; no amino acid (= stop codon)
    c. UAA; AUU; no amino acid (= stop codon)
    d. CGG; GCC; alanine
    e. CGG; ATT; isoleucine

9. Refer to Figure 13.10 in your textbook. If the highlighted triplet DNA base CTC had mutated into CTT

instead of CAC as shown, what would have been the consequence for the hemoglobin formed from this alternative mutation?
   a. Normal hemoglobin would have been made.
   b. Sickle-cell hemoglobin would have been made.
   c. Threonine would have been substituted for glutamate instead of valine.
   d. Methionine would have been substituted for glutamate instead of valine.
   e. none of the above

10. Prior to mutation, a sequence of DNA reads GAGCCTATGCCAGTA. After the mutation, the sequence reads GAGCGTACGCCATTA. Which of the following best explains the change in DNA that has occurred?
    a. There was a single base deletion.
    b. There was a single base substitution.
    c. There were multiple base deletions.
    d. There were multiple base substitutions.
    e. There were multiple base insertions.

11. In the Biology in the News box "Injections Temporarily Turn Slacker Monkeys Into Model Workers," which of the following strategies would result in a genetically engineered monkey that would play the computer game tirelessly?
    a. introducing a point mutation into the monkey's *D2* (dopamine receptor) gene, which results in an early stop codon
    b. deleting a large portion of the *D2* gene
    c. introducing a point mutation that creates a frameshift mutation that disrupts the sequence of the *D2* gene
    d. all of the above
    e. none of the above

Related Activities

- Cut out 24 half-inch squares of paper and label six of them A, six T, six C, and six G. These represent DNA bases. Cut out another 24 squares and label six more A, six U, six C, and six G. These represent RNA bases. Now randomly arrange the DNA bases into a linear sequence. What is the corresponding mRNA that would be manufactured during transcription of this DNA? Once you have completed this task, rearrange the DNA and make new mRNA as many times as necessary until you are convinced as to how transcription operates within the cell nucleus.
- After you have completed the activity above, make a second set of RNA bases. These will represent tRNA. Again, make a sequence of DNA bases and the corresponding mRNA. Now indicate the tRNA anticodons that would be needed to match the mRNA transcript. Refer to the genetic code in Figure 13.6 and write down the amino acid sequence that this segment of DNA codes for. Repeat the process until you are comfortable with converting DNA and mRNA information into an amino acid sequence, as occurs during the process of translation.
- Consider the section "From Gene to Protein, to New Hope for Huntington Disease" at the end of Chapter 13. This section highlights the discovery of the genetic basis for Huntington and other neurological disorders. The discovery that GTC repeats may lead to the production of glutamine clumps in the nerve cells of individuals has led scientists to begin targeting the prevention of clump formation as a possible cure. Search the World Wide Web for information on another neurological disorder with late age onset, Alzheimer disease. Compose a one-page comparison between the glutamine clumps discussed in the chapter and the formation of neurofibrillary tangles associated with Alzheimer disease. Does your summary indicate the potential of discovering a genetically-based cure for Alzheimer disease?

Answers and Explanations

Factual Knowledge

1. **b** Protein synthesis is a two-step process involving transcription in the cell nucleus followed by translation in the cytoplasm. All the other choices are either factually inaccurate or, if accurate, need not necessarily take place for protein synthesis to occur.
2. **e** All of the molecules listed are types of proteins, which are coded for by genes.
3. **c** This is the functional definition of one certain type of gene.
4. **a** The first step in the transcription process is the binding of RNA polymerase, the primary enzyme responsible for the process, to a specific sequence that signals the start of the gene called the *promoter*.
5. **b** The pairing of DNA nucleotides with RNA nucleotides occurs during transcription, not translation.
6. **c** Recall that during the process of transcription, RNA polymerase reads information present in the template strand of DNA, producing a complementary sequence of nucleotides in the mRNA molecule. Therefore, the sequence of the mRNA molecule duplicates the DNA sequence of the other strand, termed the *coding strand*.

7. **True** By having these two binding sites, tRNA can interact with both a specific amino acid and a unique codon on a strand of mRNA that is being read by a ribosome.

8. **b** The triangle represents the amino acid methionine, carried by the tRNA that binds to the start codon.

9. **b** Since there are two attachment sites on the ribosome, two mRNA codons (and therefore two tRNA anticodons) may be present at any one time. Review Figure 13.8 to visualize the spatial configuration of the mRNA/ribosome interaction.

10. **a** Deletion mutations of the sort described here typically result in a frameshift that ripples through the remainder of the base sequences. The result is usually an incomplete or malfunctioning protein.

11. **False** In genes that code for rRNA and tRNA, the RNA is actually the end product, and therefore does not undergo translation into a protein. Only mRNA gets translated.

12. **b** codon
    e mRNA
    c transcription
    a translation
    d intron

Conceptual Understanding

1. **c** The available evidence indicates that both DNA and RNA are present in the sample. The logic here is that thymine is unique to DNA, and both ribose and uracil are found only in RNA. With this evidence, however, you can say nothing about the types of RNA present in the sample, or whether the sample is actually from the killer.

2. **True** Both of these researchers made the link between genes and enzymes (types of proteins) through studies of inherited metabolic disorders in humans.

3. **a** RNA polymerase is the enzyme that binds to the promoter site on DNA to begin the process of gene transcription.

4. **b** The first four sets of triplet bases code for the amino acids asparagine, arginine, serine, and arginine, respectively. The last codon is a termination (stop) signal. Thus, the polypeptide coded for by this segment of mRNA will have a total of four amino acids.

5. **c** A DNA triplet of CGT in the template strand would yield an mRNA codon of GCA. Therefore, the complementary tRNA anticodon would be CGU.

6. **b** See Figure 13.3 for a pictorial representation of this information.

7. **d** A functional strand of mRNA must have both a start and a stop codon. The start codon typically codes for the amino acid methionine, which may or may not end up being a part of the final protein. However, the stop codon would not code for an amino acid. Thus, with 66 codons in the mRNA, there could be as many as 65 amino acids in the final protein product.

8. **c** The complementary base-pairing rules of DNA-to-RNA and RNA-to-RNA give the codon and anticodon. Recall also that there is no thymine in RNA; uracil and adenine are complementary bases for RNA. Figure 13.6 indicates that the mRNA codon UAA is a stop signal and does not code for any amino acid.

9. **a** If the alternative mutation had occurred, the new mRNA codon would be GAA. Using Figure 13.6, we see that GAA also codes for glutamate, just like the original DNA message, and therefore normal hemoglobin would have been produced.

10. **d** Since the total number of bases is the same, it is unlikely that either deletions or insertions occurred. In this case, however, simply comparing each strand base-for-base shows that the more likely explanation is that multiple substitutions took place along the sequence.

11. **d** All of these mechanisms would disrupt the *D2* gene and consequently the production of the dopamine receptor. This would be equivalent to the experiment discussed in the article, resulting in "model monkey workers."

CHAPTER 14

Control of Gene Expression

Getting Started

Below are a few questions to consider prior to reading Chapter 14. These questions will help guide your exploration and assist you in identifying some of the key concepts presented in this chapter.

1. Is it possible that the legendary single-eyed Greek adversary Cyclops actually existed?
2. In terms of total number of base pairs, which organisms have the largest genome?
3. What are transposons? Where did they come from, and how much of the human genome is made up of transposons?
4. If the total amount of DNA present in your body were to be stretched end to end, how long would these molecules measure, and how does this compare to the distance between the sun and Earth?
5. How many different types of cells are present in the human body? How many of these types produce the protein hemoglobin?
6. What are housekeeping genes?
7. In terms of homeotic genes, how are humans similar to fruit flies?
8. In the bacteria *E. coli*, how is the production of the amino acid tryptophan regulated?
9. What is a DNA chip, and how can it be used in the diagnosis of disease?

A Guide to the Reading

The following concepts typically give students the most difficulty when exploring the content in Chapter 14 for the first time. For each concept, one or more references have been identified that may help you gain a better understanding of these potentially problematic areas.

Transposons

The human genome consists of approximately 3.3 billion base pairs containing roughly 25,000 genes. It is estimated that only 1.5 percent of the genome contains information encoding a protein. In addition to noncoding and spacer DNA sequences, a typical eukaryotic genome also contains a large number of transposons. Transposons are sequences of DNA that are capable of changing their position within the genome. As such, they are often referred to as "jumping genes." As discussed in the text, it is estimated that upward of 36 percent of the human genome is made up of transposons. Scientists believe that transposons likely originated from a viral source. How do these jumping genes change position within the genome? In many cases, the gene sequence present in the transposon codes for an enzyme that facilitates the removal of the transposon from its original location and subsequent insertion in its new location. Recall from Chapter 12 that specialized enzymes used in the repair of DNA are capable of performing these same removal and insertion functions.

For more information on this concept, be sure to focus on:

- In Section 14.1, Genes constitute only a small percentage of the DNA in eukaryotes
- Figure 14.3, The Composition of Eukaryotic DNA
- Table 14.1, Types of Eukaryotic DNA

Differential Gene Expression

Humans have more than 200 different types of cells. As discussed in Chapter 11, each of these cells, regardless of type, contain the exact same set of genes. So how does the body produce over 200 different cell types even though the DNA present within each cell type is exactly the same? This is accomplished through the selective expression of a subset of genes required for each particular cell type. For example, skin and hair cells produce a strong, fibrous protein called *keratin* that allows these cells to perform their protective function. As mentioned in the chapter, red blood cells in the body are the exclusive producers of the oxygen-transporting protein hemoglobin. Examples such as these illustrate how cells become specialized depending upon the subset of genes that they express. In addition to this differential gene expression, a subset of genes called "housekeeping genes" is expressed in almost all cells in the body. These genes code for universally required products such as different types of RNA.

For more information on this concept, be sure to focus on:

- In Section 14.3, Different cells in eukaryotes express different genes
- Figure 14.6, Different Types of Cells Express Different Genes

Controlling Gene Expression

As discussed in the chapter, the cell uses several distinct methods for controlling the expression of genes. The primary mechanism utilized by most cells, however, is controlling the process of transcription. Regulatory DNA sequences may be used by the cell to either promote or inhibit the transcription of particular genes. Inhibition of the expression of a gene sequence can be accomplished through the binding of a repressor protein, such as that used by the tryptophan operator discussed in the chapter, to the regulatory sequence. The presence of the repressor protein can then block the binding of RNA polymerase and subsequently inhibit the process of transcription. Conversely, promoter proteins that promote the attachment of RNA polymerase may bind to regulatory sequences, which results in the enhanced transcription of the genes. To add more complexity to the process of gene regulation, it is important to understand that the activity of these promoter or repressor proteins may be turned on or off depending upon cellular or environmental conditions. Therefore, the regulation of gene expression can be thought of as an elegantly complex cascade of events that work together to make the production of proteins a highly efficient process.

For more information on this concept, be sure to focus on:

- In Section 14.4, Cells control the expression of most genes by controlling transcription
- In Section 14.4, Cells also control gene expression in other ways

- Figure 14.9, Repressor Proteins Turn Genes Off
- Figure 14.10, Control of Gene Expression in Eukaryotes

Tying It All Together

Several concepts presented in this chapter build upon concepts presented in previous chapters and may also be revisited and discussed in greater detail in subsequent chapters, including:

Environmental Influence on Gene Expression (Phenotype)

- Chapter 10 – in Section 10.5, The environment can alter the effects of a gene

Chromosomes

- Chapter 11 – Section 11.1, The Role of Chromosomes in Inheritance

One Gene – One Protein (or RNA molecule)

- Chapter 13 – Section 13.1, How Genes Work

Practice Questions

Factual Knowledge

1. Which of the following would you *not* expect to find in prokaryotic DNA?
   a. millions of base pairs
   b. histone spools around which the DNA is coiled
   c. functionally related genes grouped together in the same section of DNA
   d. a majority of DNA that codes for protein or RNA
   e. very few introns

2. Which of the following cell types found in your body would contain a different set of genes?
   a. acinar cell
   b. nerve cell
   c. epithelial cell
   d. white blood cell
   e. none of the above

3. The combined length of all of the DNA present in a single cell of your body is approximately
   a. 2 meters.
   b. the length of a football field.
   c. 10 kilometers.
   d. 100 kilometers.
   e. over 100 times the distance from Earth to the sun.

4. Which of the following would you expect to find in a sample of DNA from one of your liver cells?
   a. transposons
   b. spacer DNA
   c. DNA that codes for proteins
   d. DNA that codes for RNA
   e. all of the above

5. In eukaryotic cells, histones
   a. are the equivalent of "jumping genes."
   b. are sequences of DNA that separate genes.
   c. serve as structures around which DNA is spooled for packaging.
   d. are sequences of DNA that are removed following transcription.
   e. regulate gene expression.

6. In bacteria, genes are typically turned on and off in direct response to short-term changes caused by environmental conditions. (True or False)

7. The homeotic genes of eukaryotes
   a. are involved in the regulation of housekeeping genes.
   b. serve as "master switches" that help regulate organismal development.
   c. are unique depending upon the type of organism.
   d. probably evolved very late in the evolutionary history of organisms.
   e. all of the above

8. Which of the following represents a type of housekeeping gene that would likely be expressed in a human pancreas cell?
   a. rRNA gene
   b. hemoglobin gene
   c. insulin gene
   d. crystallin protein gene
   e. transposons

9. The most common way gene expression is regulated in both prokaryotes and eukaryotes is through the
   a. control of mRNA translation.
   b. breakdown of proteins formed by translation.
   c. prevention of DNA uncoiling prior to transcription.
   d. control of gene transcription.
   e. none of the above

10. Which of the following does *not* play a role in the genetic regulation of tryptophan synthesis in the bacterium *E. coli*?
    a. RNA polymerase
    b. homeotic genes
    c. repressor proteins
    d. operator
    e. regulatory DNA sequences

11. In addition to controlling transcription, eukaryotic cells can regulate gene expression by failing to unpack coiled DNA or influencing the speed with which mRNA is broken down prior to translation. (True or False)

12. Which of the following statements regarding the control of gene expression and phenotype is *false*?
    a. The control of gene expression influences phenotype.
    b. Since homeotic genes are always expressed, they have little effect on phenotype.
    c. The environment influences the control of gene expression and, therefore, phenotype.
    d. Phenotype is determined by a combination of genotype and the control of gene expression.
    e. none of the above

13. Match each term with the best description.
    __ spacer DNA
    __ homeotic genes
    __ operator
    __ genome
    __ gGene cascade
    a. all the DNA in an organism, including its genes
    b. controls transcription of a gene or group of genes
    c. stretches of DNA that separate genes
    d. gene products trigger expression of other genes
    e. controls a series of genes involved in development

14. Which of the following is *most* abundant in human DNA?
    a. regulatory genes
    b. noncoding DNA
    c. transposons
    d. homeotic genes
    e. rRNA genes

Conceptual Understanding

1. Consider the cyclopean organisms discussed at the beginning of this chapter. Which of the following is (are) plausible genetic explanations for the development of this condition?
   a. The organisms possessed a defective homeotic gene for eye development.
   b. The organisms lacked the genes that code for eye formation.
   c. Chemical exposure during development prevented normal eye gene products from functioning properly.
   d. both a and b
   e. both a and c

2. Refer to Figure 14.4 in your textbook. The structure labeled "metaphase chromosome" at the top of this diagram would not be able to undergo successful gene transcription because
   a. there is no operator present.
   b. the structure lacks regulatory genes.
   c. the DNA is packed too tightly.
   d. there are no homeotic genes in this structure.
   e. an active repressor is present to prevent transcription.

3. A sample of DNA from an unknown organism is analyzed and found to contain histone proteins, more than 2 billion base pairs, and large segments of noncoding DNA. From this information, one can conclude that the organism is
   a. a bacterium.
   b. an animal.
   c. a plant.
   d. some kind of eukaryotic organism.
   e. none of the above

4. From an evolutionary viewpoint, the information depicted in Figure 14.8 of your textbook about homeotic genes in mice and fruit flies suggests that
   a. these organisms are very closely related.
   b. homeotic genes differ greatly between these two animals.
   c. homeotic genes are ancient and conserved among different organisms.

d. homeotic genes on the chromosome are arranged in reverse order in the two organisms.
e. homeotic genes have little to do with controlling development.

5. Which of the following represents a situation in which human cells would likely control gene expression through the use of a gene cascade?
   a. protecting proteins against damage due to high temperatures
   b. moderating short-term changes in blood sugar
   c. making adjustments for fluctuations in blood pH
   d. coordinating critical events in embryonic development
   e. none of the above

6. In which of the following cells from a human would you expect to find the genes that code for both insulin and rRNA expressed at the same time?
   a. red blood cell
   b. eye lens cell
   c. pancreatic cell
   d. white blood cell
   e. all of the above

7. What is the relationship between homeotic genes and gene cascades?
   a. Homeotic genes regulate gene cascades.
   b. Gene cascades regulate homeotic genes.
   c. Homeotic genes respond in the short-term; gene cascades respond in the long-term.
   d. Homeotic genes are found in eukaryotes; gene cascades are found in prokaryotes.
   e. The two are essentially unrelated.

8. An *E. coli* bacterium is lacking in the amino acid tryptophan. Which of the following would be *true* about this bacterium?
   a. The tryptophan-repressor protein complex is activated.
   b. The tyrptophan-repressor protein complex can bind to the operator.
   c. Most of the repressor protein is bound to tryptophan, forming the tryptophan-repressor protein complex.
   d. RNA polymerase can bind to the operator.
   e. Enzymes for tryptophan synthesis are not being manufactured.

9. The regulation of tryptophan synthesis in *E. coli* is an example of affecting gene expression through
   a. translational control.
   b. transcriptional control.
   c. destruction of completed enzymes involved in tryptophan synthesis.
   d. breaking down mRNA molecules.
   e. the use of transposons.

10. To control the translation of gene products in both prokaryotes and eukaryotes, regulatory proteins must interact with their appropriate regulatory DNA sequences to switch these genes on or off. (True or False)

11. A certain mutation in *E. coli* causes production of a faulty repressor protein used in the control of tryptophan synthesis. Refer to Figure 14.9 in your textbook and determine the most likely effect of this mutation.
   a. The cell will make new tryptophan only when tryptophan levels are low.
   b. High tryptophan concentrations will cause the synthesis system to shut down.
   c. The cell will make new tryptophan regardless of tryptophan concentration.
   d. The ability of RNA polymerase to bind to the operator will decrease.
   e. There will be no noticeable effect on tryptophan synthesis.

12. Which of the following findings derived from recent analysis of the human genome does *not* illustrate our genetic relationships to more “primitive” organisms?
   a. Only 25,000 genes are required to make a human.
   b. Human DNA contains hundreds of bacterial genes.
   c. Numerous homeotic genes are shared among humans and other animals.
   d. Scientists have discovered that the use of DNA chips can help to identify genes expressed in cancerous cells.
   e. All of the above are true.

13. Which of the following statements regarding the value of DNA chips is *false*?
   a. Researchers can now determine how an entire genome functions under particular conditions.
   b. DNA can be monitored for the presence of particular mutations.
   c. The activity of genes in different organisms can be compared using the same chip.
   d. DNA chips can serve as a screening tool for certain genetic diseases.
   e. none of the above

Related Activities

- Recall a discussion from earlier chapters about the taxonomic and evolutionary criteria used to classify organisms. Now reflect on the commonality of homeotic genes among organisms. Combining these two sets of ideas, explain why it is legitimate for scientists to claim that homeotic genes must have originated early in the history of life on Earth.

- Review the Science Toolkit section at the end of the chapter; it discusses the use of DNA chip technology to monitor gene expression. Search the World Wide Web for information on different medical applications of this technology. Identify a unique use of this genetic technique, and compose a one-page essay explaining how the technique is being used to achieve some improvement in human medicine.
- Consider the Biology in the News story at the end of this chapter about RNA interference (RNAi) experiments lowering cholesterol levels in mice. Search the World Wide Web for information on diseases caused by the detrimental production, or overproduction, of a particular gene or protein product. Describe how RNAi may potentially be used to develop a treatment for your identified disorder.

Answers and Explanations

Factual Knowledge

1. **b** Histone proteins are used to coil the large amount of DNA found in eukaryotes so that it can be packaged more efficiently within the cell. Prokaryotes do not face this problem because they have much less DNA.
2. **e** Recall that in an individual organism, each cell would contain the exact same set of chromosomes, and therefore the same set of genes.
3. **a** The amount of DNA packed within a human cell is approximately 2 meters in length. If all of the DNA in a human body were extended end-to-end, the distance would roughly equal 130 times the distance from the Earth to the sun.
4. **e** Liver cells are representative of your body cells in general, and therefore all of these DNA features would be expected to exist.
5. **c** Recall that DNA in eukaryotes is highly compacted. This is accomplished through the formation of histone spools in which DNA is wound around a complex of histone proteins. Review Figure 14.4 for more information.
6. **True** Most prokaryotes are not long-lived, and so short-term responses to environmental changes, such as nutrient supply, are the main focus of their gene activity.
7. **b** Recall that homeotic genes play a central role in controlling gene cascades. Because of this central role and the fact that these genes are similar between groups of organisms, homeotic genes are believed to be ancient in their origin.
8. **a** Transposons are DNA sequences that are capable of changing position within a chromosome. The hemoglobin, insulin, and crystallin protein genes are specialty genes that are expressed only in cells that need those products. rRNA genes are needed in all cells that carry out protein synthesis, and so qualify as housekeeping genes.
9. **d** All of these are ways that gene expression can be regulated, but transcriptional control is clearly the most common mechanism.
10. **b** Homeotic genes are used by eukaryotes to regulate the interaction among groups of genes in long-term, complex processes such as development.
11. **True** See Figure 14.10 for this information.
12. **b** Errors in the expression of homeotic genes during development can have dramatic effects on the phenotype of an organism.
13. **c** spacer DNA
    e homeotic gene
    b operator
    a genome
    d gene cascade
14. **b** Only 1.5 percent of the human genome encodes for a protein, rRNA or tRNA. Therefore, the majority of human DNA consists of noncoding DNA sequences.

Conceptual Understanding

1. **e** Both a and c are plausible explanations for the development of a cyclopean organism. Answer b is not possible because one eye does develop, which would not be possible if the genes for eye formation are completely absent.
2. **c** Such tightly packed DNA will not undergo transcription because the necessary enzymes that control transcription cannot gain access to the DNA.
3. **d** These three clues are all characteristics of eukaryotic DNA, but we have no information that permits us to decide from which kingdom of eukaryotes the sample came.
4. **c** Since these genes are in the same locations on the chromosomes and perform the same functions in the two unrelated animals, the reasonable conclusion is that they must have appeared early on in, and been conserved throughout, the evolution of the animal kingdom.

5. **d** Homeotic genes tend to mediate long-term processes such as development, a process that involves the interactions of many different genes through gene cascades.

6. **c** Recall that rRNA genes are considered to be housekeeping genes; therefore they should be expressed in all cell types. The product insulin is differentially expressed in the cells of the pancreas.

7. **a** Gene cascades are complex interactions between gene products that trigger the response of additional genes. Homeotic genes help control these cascades.

8. **d** Since tryptophan is lacking, it would not be present in quantities sufficient to form the active tryptophan-repressor protein complex. RNA polymerase is thus not prevented from binding to the operator. This in turn causes the transcription of mRNA that codes for the production of the enzymes used in tryptophan synthesis.

9. **b** The presence or absence of tryptophan determines whether the genes that code for the necessary enzymes in tryptophan synthesis will even be transcribed.

10. **False** The word *translation* is the key here. Regulatory proteins and DNA sequences are part of transcription, not translation.

11. **c** Since the repressor is faulty, it cannot bind to the operator and prevent RNA polymerase from transcribing the genes used in tryptophan synthesis. Thus, even if tryptophan is present in the cell, the normal repression system for these genes cannot function, and new tryptophan will be made regardless of tryptophan levels.

12. **d** The fact that certain genes are expressed only in cancer cells as revealed through DNA chip analysis is not indicative of the genetic relationships between organisms.

13. **c** Refer to the section "From Gene Expression to Cancer Treatment" and the Science Toolkit section at the end of the chapter for a discussion of the scientific and medical utility of DNA chip technology. DNA chips are typically used to study the expression of genes in a single organism or cell type under different conditions.

CHAPTER 15 DNA Technology

Getting Started

Below are a few questions to consider prior to reading Chapter 15. These questions will help guide your exploration and assist you in identifying some of the key concepts presented in this chapter.

1. How was Alba, the white rabbit, made to glow in the dark?
2. What is the natural function of restriction enzymes?
3. Why must DNA be heated during the course of a DNA hybridization experiment?
4. What would you expect to find in a DNA library?
5. What are the advantages of using the polymerase chain reaction (PCR) for gene cloning?
6. Why would DNA fingerprinting not be useful to distinguish between identical twins?
7. What percentage of the world's soybeans, corn, cotton, and canola crops was genetically modified in 2002?
8. How did gene therapy help Ashanthi DeSilva to lead a normal life?

A Guide to the Reading

The following concepts typically give students the most difficulty when exploring the content in Chapter 15 for the first time. For each concept, one or more references have been identified that may help you gain a better understanding of these potentially problematic areas.

Restriction Enzymes and Gel Electrophoresis

Restriction enzymes are capable of breaking the covalent bonds that hold the long, individual strands of DNA together. What makes restriction enzymes useful for the analysis of DNA is the specificity they display in selecting a region in which to cut the DNA molecule. Each restriction enzyme has a target sequence, a specific sequence of nucleotides typically 10 base pairs or fewer in length, which must be present in order for the enzyme to be able to attach to the DNA molecule and cut the two strands. This target sequence varies from enzyme to enzyme (see Figure 15.1). Since identical copies of a DNA molecule (i.e., chromosome) would have the same set of target sequences, a restriction enzyme would cut these copies in the exact same location. However, if the copies of DNA under analysis were different in their nucleotide sequences, a single restriction enzyme would recognize and cut only those DNA molecules with the correct target sequence, resulting in DNA fragments of differing sizes. In a test tube, it would be impossible to visualize these DNA fragments. Therefore, scientists have devised a way to separate these DNA fragments based on size using gel electrophoresis. In gel electrophoresis, the DNA samples are loaded onto a slab of a Jell-o like matrix and subjected to an electrical charge. The DNA fragments, which are negatively charged, are attracted to the positively charged end of the gel slab and slowly migrate in that direction. Larger fragments move more slowly since they have a more difficult time making their way through the matrix while smaller fragments move more rapidly. Fragments of the same size tend to move as a group, accumulating in a region called a *band*. By analyzing these separated bands, scientists can determine whether or not a sample of DNA contained one or more target sequences for a particular restriction enzyme. This technique is particularly useful as a DNA fingerprinting tool.

For more information on this concept, be sure to focus on:

- In Section 15.1, Several key enzymes are used in DNA technology
- In Section 15.1, Gel electrophoresis sorts DNA fragments by size
- Figure 15.1, Restriction Enzymes Cut DNA at Specific Places
- Figure 15.2, Gel Electrophoresis
- Figure 15.3, Identifying the Sickle-Cell Allele with Restriction Enzymes and Gel Electrophoresis

Polymerase Chain Reaction

The polymerase chain reaction (PCR) is a technique that utilizes the enzyme DNA polymerase in a test tube to produce millions of copies of a piece of DNA in a short period of time. The key to understanding PCR is realizing that a portion of the sequence of DNA that a scientist wishes to copy (or amplify) must already be known. This is because the DNA polymerase enzyme is only capable of extending a DNA strand; it cannot build a DNA strand from scratch. Therefore, PCR requires the use of *primers* (see Figure 15.8), which are short segments of DNA that match the two ends of the sequence of DNA targeted for copying. In a typical PCR reaction, the two strands of template DNA are separated using heat. By cooling the reaction a little, the primer DNA strands are allowed to bind to their complementary sequences located on the target strands. DNA polymerase can then begin to extend the primer using the original DNA as a template. This results in two identical copies of the target DNA. This cycle is repeated 20–30 times, each time doubling the number of copies of the target DNA in the test tube, until billions of copies of the DNA sequence of interest are obtained.

For more information on this concept, be sure to focus on:

- In Section 15.2, The polymerase chain reaction can clone DNA very rapidly
- Figure 15.8, The Polymerase Chain Reaction

Cloning

As presented in the chapter, cloning can mean several things. In DNA cloning, a single copy of a gene is isolated and inserted into some type of vector for easy copying and subsequent transfer to another organism. On a larger scale, the cloning of entire organisms is referred to as reproductive cloning, while the cloning of tissues or organs is termed therapeutic cloning. If you are familiar with the story of Dolly the sheep, you are aware of the process of reproductive cloning, where an offspring that is an exact genetic duplicate of an existing individual is produced. The process involves the transfer of a diploid nucleus isolated from the cell of an individual into an egg cell from which the nucleus has been removed. In this case, the imported diploid nucleus provides the entire genetic blueprint for the development of the egg into an embryo, a fetus, and ultimately a newborn. Since the DNA present in the newborn comes entirely from an existing individual, the newborn is genetically identical to the DNA donor and therefore represents a clone of this individual. In therapeutic cloning, stem cells are harvested from the developing embryo. In this case, these genetically identical stem cells may then be implanted or transferred to the DNA donor to treat a particular disorder, such as a degenerative neurological disease, or to restore the normal function of a tissue. The key to both these types of cloning procedures is the fact that the resulting stem cells or organisms are all genetic duplicates of the original donor.

For more information on this concept, be sure to focus on:

- Section 15.2, Working with DNA: DNA Cloning
- Chapter 15 – Science Toolkit, Human Cloning

Tying It All Together

Several concepts presented in this chapter build upon concepts presented in previous chapters and may also be revisited and discussed in greater detail in subsequent chapters, including:

DNA Structure and Replication

- Chapter 12 – Section 12.3, How DNA Is Replicated

Single Gene Mutations and Disease

- Chapter 13 – Section 13.6, The Effect of Mutations on Protein Synthesis

Practice Questions

Factual Knowledge

1. Which of the following are products of genetic engineering?
   a. establishment of dog breeds
   b. agricultural crops with enhanced nutritional value
   c. DNA fingerprinting
   d. RFLPs
   e. all of the above

2. The primary reason why the same basic techniques can be used to analyze the DNA from species as diverse as bacteria and humans is that
   a. all cells are identical.
   b. every organism has the same amount of DNA.
   c. the DNA sequences of all organisms are the same.
   d. DNA has a consistent structure in all organisms.
   e. all of the above

3. To study the genetic basis of the inherited human disease sickle-cell anemia, a biologist first isolates DNA from an affected individual's cells. Which of the following would be the third step in this process?
   a. Use restriction enzymes to break the DNA into smaller fragments.
   b. Apply electrical current across the gel to separate DNA fragments.
   c. Visualize DNA fragments in the gel using a stain.
   d. Load DNA fragments onto a gel.
   e. Use polymerase chain reaction to produce many copies of the target DNA sequence.

4. Which of the following statements about restriction enzymes is *false*?
   a. They work on DNA from all types of organisms.
   b. They are used to glue together short segments of DNA.
   c. They come in many varieties, each with its own DNA target sequence.

d. They are highly specific for their DNA target sequences.
e. They are believed to be a primitive type of immune system used by bacteria to degrade viral DNA.

5. DNA probes are
a. large segments of double-stranded DNA.
b. enzymes used in the polymerase chain reaction.
c. short, single-stranded segments of DNA used in DNA hybridization experiments.
d. used to generate restriction fragment length polymorphisms.
e. none of the above

6. In gel electrophoresis, the distance traveled by a DNA fragment
a. corresponds to the size of the fragment.
b. depends on which ligase was used to make the fragments.
c. is related to the species of organism from which the DNA came.
d. depends on whether or not the fragment is attached to a plasmid.
e. all of the above

7. Which of the following is a critically important tool used in experiments involving DNA hybridization?
a. restriction enzymes
b. DNA probes
c. heat
d. radioactive or chemical labels
e. all of the above

8. A gene is said to be cloned if
a. the DNA sequence of the gene is known.
b. the function of the gene is known.
c. there is a DNA probe for the gene.
d. the gene has been isolated and copied.
e. it has been used by scientists for gene therapy.

9. DNA sequencing is an integral step in the formation of DNA libraries. (True or False)

10. Which of the following is *not* necessary to execute a successful polymerase chain reaction?
a. all four DNA bases
b. short single-stranded DNA primers
c. DNA polymerase
d. DNA library
e. a target DNA sequence

11. In genetic engineering, genes can be inserted from one organism into another or back into the original organism using which of the following techniques?
a. polymerase chain reaction
b. gene gun
c. DNA hybridization
d. gel electrophoresis
e. gene therapy

12. Which of the following DNA sources would be suitable for DNA fingerprinting analysis?
a. blood
b. skin
c. semen
d. saliva
e. all of the above

13. Which of the following statements regarding DNA technology is *false*?
a. DNA technology can be used to alter the performance of a modified organism.
b. DNA technology can be used to produce multiple copies of a desired DNA sequence.
c. DNA technology can be used to generate many copies of specific genes.
d. DNA technology can enhance the production of a specific gene product.
e. DNA technology can be used to remove undesired alleles from a multicellular organism.

14. In human gene therapy, doctors
a. seek to correct genetic disorders by repairing the genes that cause them.
b. treat the symptoms of a disorder, but not the underlying genetic cause.
c. must work at the embryonic stage of development to be effective.
d. know with certainty that the therapy alone improves the patient's condition.
e. have been able to use the viral method to deliver modified genes.

15. Match each term with the best description.
__ DNA probe
__ clone
__ plasmid
__ DNA library
__ gene therapy
a. a vector for DNA transfer between cells
b. effort to fix the genes that cause disorders
c. DNA sequence that can pair with a particular gene
d. one organism's DNA stored in a host organism
e. isolated gene of which many copies are made

Conceptual Understanding

1. The gene that controls the production of the firefly enzyme luciferase can be inserted into tuberculosis (TB) bacteria. This technology provides a human medical benefit because physicians can now more easily determine

a. whether or not the patient has TB.
b. where the TB infection came from.
c. if the strain of TB being tested is resistant to antibiotics.
d. if the TB infection is related to tobacco use.
e. all of the above

2. Imagine that, sometime in the future, human space explorers discover life on another planet and that this life has DNA as its genetic material. Which of the following could be used to study the genetics of this extraterrestrial life?
   a. cloning
   b. gel electrophoresis
   c. polymerase chain reactions
   d. restriction enzymes
   e. all of the above

3. DNA ligases and polymerases are key components of DNA technology because they help break down the organismal genome into DNA fragments. (True or False)

4. Imagine a gel through which DNA fragments have moved in response to an applied electrical current. The band on this gel that is farthest from the top (that is, from the place where the DNA fragments were added to the "well") represents the
   a. shortest fragments of DNA.
   b. longest fragments of DNA.
   c. restriction enzyme used to cut the DNA into fragments.
   d. ligase used to bind the DNA fragments together.
   e. DNA polymerase used to make copies of the DNA fragments.

5. Refer to Figure 15.3 in your textbook. At the bottom of the diagram is the result of a gel electrophoresis run using the DNA fragments from a sickle-cell allele and a normal allele that were generated using the restriction enzyme *DdeI*. Which of the fragments (x, y, or z) shown is the largest?
   a. x
   b. y
   c. z
   d. They are all the same size.
   e. cannot determine from the information given

6. A sample of DNA from a person suspected of having sickle-cell anemia is subjected to DNA hybridization using two probes—one that binds to the normal allele and another that binds to the sickle-cell allele. If both probes bind to the DNA, this individual
   a. is homozygous for the sickle-cell gene.
   b. is heterozygous for the sickle-cell gene.
   c. is normal.
   d. has sickle-cell anemia.
   e. none of the above

7. DNA cloning is beneficial because a cloned gene can be
   a. sequenced using automated sequencing machines.
   b. transferred to other cells of the same organism.
   c. transferred to cells in other organisms.
   d. used in DNA hybridization experiments.
   e. all of the above

8. Consider textbook Figure 15.7. What would be the effect of using one type of restriction enzyme to cut the organism's DNA, and then a different one to cut the plasmid DNA in the first step?
   a. Incompatible DNA fragments would be produced.
   b. A different ligase would be needed to join the DNA fragments.
   c. A different kind of plasmid would be needed as a vector.
   d. Bacteria would not be able to take up the plasmids produced.
   e. The results would be the same as those pictured in the figure.

9. A biologist intends to use a polymerase chain reaction to perform a task. The biologist probably is trying to
   a. discover new genes.
   b. clone a gene.
   c. cut DNA into many small fragments.
   d. isolate DNA from a living cell.
   e. separate DNA fragments from one another.

10. In a criminal trial in which DNA fingerprinting has been used, a genetic "match" between the suspect and evidence left at the crime scene provides
    a. definitive proof that the suspect is guilty.
    b. a high probability that the suspect is guilty.
    c. a low probability that the suspect is guilty.
    d. definitive proof that the suspect is innocent.
    e. no useful information.

11. Prenatal screening in humans, AIDS virus testing in humans, and genetic engineering for protection against insect attack in plants all
    a. are types of gene therapy.
    b. are examples of the application of DNA technology.
    c. require use of the same restriction enzyme.
    d. are carried out using gene guns.
    e. involve no ethical dilemmas or environmental risks.

12. Genetically engineered bovine growth hormone (BGH), which is highly effective for improving overall growth and milk production in cattle, remains a hotly debated issue because
    a. BGH is clearly hazardous to human health.
    b. BGH is an environmental hazard.
    c. BGH could drive traditional family farmers out of business.
    d. scientists remain unconvinced that BGH really works.
    e. all of the above

13. Gene therapy has been tested on a large number of patients with a wide variety of inherited genetic disorders, and in numerous cases it has produced a complete cure. (True or False)

14. Based on the Biology Matters box, "Think Before You Drink," which of the following statements is *false*?
    a. Consumers will have the final say regarding the success of genetically modified crops.
    b. Genetically modified crops are used only for the production of breads and cereals.
    c. Genetically modified crops have allowed the United States to remain among the most agriculturally productive nations.
    d. The United States is one of the global leaders in the production of genetically modified crops.
    e. none of the above

Related Activities

- Search the World Wide Web to find information on stem cell research. Compose a one-page essay summarizing what you have learned, and indicate specifically how stem cell research may potentially be used to treat patients with various degenerative diseases.
- List three issues associated with the use of DNA technology that pose potential ethical dilemmas or societal risks. What are these social and ethical concerns? Short of altogether prohibiting the use of new DNA manipulation techniques, what can be done to adequately control or regulate work in this area?
- Review the Biology in the News article "Biotech Grass Found Far Afield" at the end of the chapter. Describe the consequences of finding genetically engineered genes present in native grasses many miles away from the experimental plot of bentgrass. Should such observations concern consumers of genetically engineered foods? Why or why not?

Answers and Explanations

Factual Knowledge

1. **b** As described in the chapter's introduction, new genes may be inserted into various food crops enabling them to produce essential nutrients that would otherwise be lacking.
2. **d** The fact that DNA is structured the same way in all known organisms means that similar methods can be used to study the hereditary material.
3. **b** Once isolated, DNA can be cut into fragments using restriction enzymes followed by separation of the fragments using gel electrophoresis. The third step in the process would involve applying the electrical charge across the gel to begin separating the DNA fragments by size.
4. **b** Recall that ligases are the enzymes used to glue together DNA fragments.
5. **c** DNA probes are used in hybridization experiments. These short, single-stranded segments of DNA are used to identify the presence of target DNA sequences.
6. **a** The determining factor in the movement of DNA through an electrophoresis gel is the size of the fragment.
7. **e** DNA hybridization involves the breaking of DNA into smaller fragments using restriction enzymes, heating the DNA to separate the strands followed by probing for target sequences using labeled DNA probes.
8. **d** Although cloned genes may be used as part of, or in association with, the other choices, the definition of a cloned gene is one that has been isolated and repeatedly duplicated.
9. **False** DNA sequencing is performed to determine the nucleotide sequence of a DNA fragment or gene. DNA libraries represent a collection of DNA fragments stored in a host organism using a type of cloning vector. DNA sequencing can be performed on a cloned vector obtained from a DNA library.
10. **d** Although a DNA library is often created prior to doing PCR, it is not necessary. All that is needed is one desired sequence of DNA, not an entire library.
11. **b** Besides the gene gun, other methods of genetic transfer between cells or organisms include the use of vectors such as viruses and plasmids.

12. **e** DNA fingerprinting can be performed with virtually any source containing DNA, including the tissues and fluids mentioned in the question.

13. **e** While DNA technology is often used to introduce foreign genes into an organism, it cannot be used to remove diseased genes from a multicellular organism since each cell in the body would need to be treated separately.

14. **a** Gene therapy, while showing great theoretical promise, is still in its infancy as a practical means of treating the cause of genetic disorders. To date, there has been limited success and a few disturbing failures with the technique, but it is important to note that the positive results have not been clearly attributable to gene therapy alone.

15. **c** DNA probe
 e clone
 a plasmid
 d DNA library
 b gene therapy

Conceptual Understanding

1. **c** Resistant strains of TB will express the firefly gene and therefore glow in the dark! This feature aids in treatment of the disease by dramatically speeding up diagnostic time for the slow-growing bacteria.

2. **e** All of these choices are used to study DNA-based genetic systems on Earth, and they would likely be universal among other DNA systems as well.

3. **False** Both ligases and polymerases are used to synthesize and assemble DNA sequences, not to break them down.

4. **a** Migration through the electrophoresis gel is a function of the size of the DNA fragments, with small fragments moving farthest as they are able to "squeeze" through the gel more easily.

5. **c** Since it moved the shortest distance through the gel, DNA sequence z must be the largest. This is why we can conclude that z represents the sickle-cell allele, which cannot be cut by the restriction enzyme used in this test. x and y are smaller fragments produced when the normal allele is cut by the enzyme.

6. **b** Since both probes bind, both of the target DNA sequences (alleles) must be present. The individual is therefore heterozygous.

7. **e** All of these are possible once a gene has been cloned.

8. **a** In order for the cut organismal and plasmid DNA fragments to connect properly, they must have similar ends on their DNA sequences following digestion with restriction enzymes. This can be accomplished only by using the same restriction enzyme. Using different enzymes for each type of DNA would result in incompatible fragments with different sequences at their ends.

9. **b** PCR is a way to dramatically increase the numbers of a particular sequence of DNA (or an entire gene). This technique is most useful when cloning DNA since prior knowledge of the DNA sequence must first be obtained.

10. **b** Although DNA fingerprinting is a powerful technique, a genetic match can demonstrate only a high probability of guilt since there is a 1/100,000 to 1/1,000,000,000 chance that two individuals will share the same genetic profile.

11. **b** These are all good examples of the end products of various DNA technologies.

12. **c** The use of BGH does not appear to have serious health or environmental drawbacks, but it could offer such a competitive edge to corporate farming operations that smaller, more traditional family farms might be forced out of business. This scenario is a good example of how genetic engineering can have unintended social consequences.

13. **False** The application of gene therapy is still quite limited, and the few apparent successes are open to speculation about just how much the gene therapy actually accounted for the patient's improvement.

14. **b** Genetically modified foods are found in many commercial products such as frozen pizzas, oils, and corn syrup.

INTERLUDE C

Harnessing the Human Genome

Getting Started

Below are a few questions to consider prior to reading Interlude C. These questions will help guide your exploration and assist you in identifying some of the key concepts presented in this interlude.

1. What scientific milestone was achieved in February of 2001?
2. What are SNPs, and how do scientists use them to predict an individual's susceptibility to a genetic disease?
3. Why did the National Institutes of Health decide to use a mixture of DNA from several individuals for the Human Genome Project?
4. How is it that a complex organism such as a human can be produced using only ten times the number of genes required to produce a simple bacterium (like *E. coli*)?
5. What have scientists learned about human sleep patterns by studying fruit flies?
6. What is behavioral genomics, and why are ethicists alarmed about this emerging field?

A Guide to the Reading

The following concepts typically give students the most difficulty when exploring the content in Interlude C for the first time. For each concept, one or more references have been identified that may help you gain a better understanding of these potentially problematic areas.

Organismal Complexity

The number of genes present in an organism's genome does not always reflect how complex the organism may be. For instance, humans have a genome size of 3.3 billion base pairs with 20–30,000 functional genes (See Table C.1). Compare this to some of the single-celled organisms listed in the table that have 5–6,000 genes in their genome. To further illustrate this point, consider that the puffer fish *Takifugu rubripes* has an estimated 31,000 genes! So how is it that organisms with such varied complexity can be produced using roughly the same number of genes? Part of this can be explained by understanding that it is the combination of genes that are turned on or off that helps to generate such diversity. As described in the interlude, an organism with 100 total genes that can be either on or off can have a total of 2^{100} possible genetic combinations. Recall that the expression of genes depends on several interacting factors. Also, eukaryotic genes may be processed (removal of introns) in such a manner that makes it possible to generate a number of different mRNA molecules from a single gene. All of these factors taken together make it easy to see how a relatively small number of genes may work together to produce a highly complex organism.

For more information on this concept, be sure to focus on:

- In Interlude C, Complexity in organisms is determined by more than just the number of genes
- Table C.1, A Sample of the Genomes Sequenced to Date
- Figure C.3, How Genes Can Produce Multiple RNAs

Functional Protein Domains

A protein domain is a portion of a protein, such as an enzyme, that has a particular function. Life processes that are common between groups of organisms, such as DNA replication and glycolysis, make use of enzymes that are likewise very similar. Since any alteration to the amino acid sequence of these important enzymes would likely have a negative impact on the function of the protein, mutation within these domains is not tolerated very well. This explains the observation that 90 percent of all human protein domains are also found in less complex organisms such as nematode worms and fruit flies. The pressures to maintain consistent sequences of genes will be discussed in more detail in Chapter 18.

For more information on this concept, be sure to focus on:

- In Interlude C, Comparing the genomes of different organisms yields valuable insights
- Figure C.4, Gene Expression in a Fruit Fly Embryo

Single Nucleotide Polymorphisms

Recall that the DNA from two unrelated individuals is approximately 99.9 percent identical in nucleotide sequence. Single nucleotide polymorphisms (SNPs) represent the single base-pair differences that make each of us unique. The Human Genome Project has identified more than 10,000,000 SNPs. SNPs can be useful for genetic screening when a particular SNP can be associated with a particular genetic disease. It is important to note that the SNP itself may not necessarily be the cause of the genetic disease, but it does provide a "marker" that can be screened for to determine an individual's predisposition to developing the associated disease. Several examples are provided in the interlude, including the identification of different SNPs within the *apoE* gene that are associated with the development of Alzheimer disease. When combined with the power of the DNA chip, SNPs provide a rapid, accurate method for doctors to obtain a person's genetic profile.

For more information on this concept, be sure to focus on:

- In Interlude C, SNPs are a powerful means of characterizing individual genomes
- In Interlude C, SNP profiling will help identify the risk of getting certain diseases
- Figure C.6, From Tissue to Test Results

Tying It All Together

Several concepts presented in this chapter build upon concepts presented in previous chapters and may also be revisited and discussed in greater detail in subsequent chapters, including:

Cancer

- Interlude B, Cancer: Cell Division Out of Control

Human Genetic Disorders

- Chapter 11 - Section 11.5, Human Genetic Disorders

Preimplantation Genetic Diagnosis

- Chapter 11 - Science Toolkit, Prenatal Genetic Screening

Noncoding DNA Sequences

- Chapter 13 - Section 13.3, Transcription: Information Flow from DNA to RNA

Gene Expression

- Chapter 14 - Section 14.4, How Cells Control Gene Expression

DNA Technologies

- Chapter 15 - Section 15.3, Applications of DNA Technology

Evolution

- Chapter 17 - Section 17.7, Natural Selection: The Effects of Advantageous Alleles

Practice Questions

Factual Knowledge

1. With which of the following organisms did scientists first obtain a complete genome draft?
   a. humans (*Homo sapiens*)
   b. yeast (*Saccharomyces cerevisiae*)
   c. fruit flies (*Drosophila melanogaster*)
   d. nematode worms (*Caenorhabditis elegans*)
   e. bacteria (*Haemophilus influenzae*)

2. The field of genomics
   a. analyzes the process of inheritance.
   b. analyzes the structure of DNA molecules.
   c. focuses on individual genes and how they function.
   d. focuses on the structure and expression of entire genomes.
   e. all of the above

3. Which of the following statements regarding single nucleotide polymorphisms (SNPs) is *false*?
   a. SNPs represent a significant source of genomic variation.
   b. Matching SNPs to a particular genetic disease requires knowledge of the DNA sequence for a particular region.
   c. Currently, only a few hundred SNPs have been identified through the Human Genome Project.
   d. SNPs can be screened readily using DNA chip technology.
   e. none of the above

4. What percentage of human proteins are shared with fruit flies?
   a. Less than 1%
   b. 10%
   c. 20%
   d. 40%
   e. 60%

5. The human genome consists of over ________ base pairs.
   a. 1 million
   b. 1.8 million
   c. 3.6 million
   d. 3.2 billion
   e. 46 billion

6. Since noncoding regions of DNA are not directly responsible for producing proteins, it is not necessary for scientists to sequence these regions in order to understand human genomics. (True or False)

7. The draft map of the human genome was completed almost five years ahead of schedule. Which of the following statements regarding the Human Genome Project is *false*?
   a. The first draft of the human genome was completed in a mere 15 years.
   b. The Human Genome Project was conducted exclusively by laboratories in the United States.
   c. The DNA used to complete the Human Genome Project was obtained from several diverse donors.
   d. Before publishing the complete draft of the human genome, the Human Genome Project published the complete genome sequence of the yeast *S. cerevisiae.*
   e. None of the above statements are false.

8. For which of the following applications would SNP profiling be useful?
   a. evaluating a patient's response to a particular drug treatment
   b. assessing a patient's genetic predisposition to a particular disease
   c. assisting physicians in prescribing a more effective health regimen for their patients
   d. providing a rapid determination of an individual's genotype for a particular gene
   e. all of the above

9. The entire human genome has been sequenced and is currently available for the entire scientific community to access. (True or False)

10. The total number of genes estimated to be present in the human genome is
    a. 1,000
    b. 10,000
    c. 30,000
    d. 100,000
    e. 3 billion

11. A protein domain is the region of a
    a. gene that codes for a protein.
    b. gene that contains regulatory sequences.
    c. protein that has a specific function.
    d. protein that is highly variable among species.
    e. none of the above

12. For which of the following diseases are genetic screening tests currently available?
    a. Huntington disease
    b. cystic fibrosis
    c. breast cancer
    d. Alzheimer disease
    e. all of the above

13. Match each term with the best description.
    __ single nucleotide polymorphism
    __ comparative genomics
    __ Human Genome Project
    __ DNA chip
    __ Celera Genomics
    a. a field of study that analyzes and compares genomes from multiple organisms
    b. massive effort designed to decipher the genetic code of *Homo sapiens*
    c. a biotechnology company that provided the first complete genome of a free-living organism
    d. a position within a gene that varies among individuals and may be associated with a genetic defect or disease
    e. a technology that may be used to generate a rapid genetic profile of an individual

Conceptual Understanding

1. Which of the following is *not* a feature that distinguishes the field of genomics from the field of genetics?
   a. Genetics analyzes the mechanisms and patterns of inheritance, whereas genomics does not.
   b. Genomics is concerned with broad-scale expression of an entire set of genes, whereas genetics deals primarily with the expression of smaller subsets of genes.
   c. Genomics makes use of DNA sequence data, whereas genetics does not.
   d. Genomics focuses on the coordinated activities of all the genes in a genome, whereas genetics focuses on individual genes and how they function.
   e. All of the above distinguish between the fields of genetics and genomics.

2. Consider an experiment conducted with cells collected from a human tumor. A SNP profile is created using DNA derived from the cancerous cells and DNA chip technology. What can such an experiment tell an investigator?
   a. whether the tumor is malignant or benign
   b. which genes have been turned "on" and which genes have been turned "off" in the cells
   c. the exact DNA sequence present in the cells
   d. the age of the patient
   e. none of the above

3. Which of the following statements regarding genetic screening is *false*?
   a. Mutations in a given gene are often associated with a genetic disease.

b. Assessing the mutations present in an individual can help determine susceptibility to developing particular diseases.
c. Prenatal screening can determine whether an *in vitro*-fertilized fetus possesses a particular genetic disease.
d. There are only a few genetic screening tests currently available.
e. none of the above

4. Which of the following would *not* be considered a valid reason for the National Institutes of Health proposal for determining the genome sequences of 12 additional species?
   a. The proposed organisms represent potential models for human disease.
   b. The proposed organisms are most closely related to humans.
   c. The proposed organisms have potential economic importance.
   d. The proposed organisms may hold the keys to identifying new genetic disorders.
   e. none of the above

5. In general, human DNA is 99.9 percent identical between individuals. Why, then, was DNA from several individuals used for the Human Genome Project?
   a. to avoid the potential of sequencing a single genome with a significant number of mutations
   b. because it was easier to sequence DNA samples from several different sources than DNA from a single source
   c. so that each lab participating in the Human Genome Project could have its own unique samples
   d. to help in generating a more detailed map of the genome
   e. none of the above

6. During the early years of the Human Genome Project, researchers focused on obtaining the sequences of smaller genomes. Which of the following was a prime reason for this approach?
   a. The smaller genomes provided an opportunity for researchers to practice obtaining genomic data.
   b. The smaller genomes allowed scientists to test new sequencing methods.
   c. The smaller genomes provided scientists with valuable practice analyzing the huge quantities of data generated during genome sequencing projects.
   d. The smaller genomes provided an opportunity for scientists to work with and develop computer programs to assist with the assembly and analysis of large amounts of data.
   e. all of the above

7. Which of the following is *not* a difference between comparative and behavioral genomics?
   a. Comparative genomics considers the genomes of multiple species whereas behavioral genomics does not.
   b. Behavioral genomics considers genetic predispositions to conditions while comparative genomics does not.
   c. Comparative genomics helps scientists to decipher evolutionary conservation across species while behavioral genomics does not.
   d. The field of behavioral genomics has societal and ethical implications while the field of comparative genomics has few or no such implications.
   e. All of the above are key differences.

8. Consider a hypothetical pair of organisms. Species A contains a total of 3 genes while Species B contains a total of 5 genes. Assuming that each gene can have 2 "states" (on or off), how many different RNA populations would be possible in each organism?
   a. Species A = 3; Species B = 5
   b. Species A = 6; Species B = 10
   c. Species A = 9; Species B = 15
   d. Species A = 8; Species B = 32
   e. Species A = 10; Species B = 64

9. Comparisons of gene sequences between humans and other organisms can
   a. reveal evolutionary conservation.
   b. yield valuable information regarding the function of particular genes.
   c. help to identify genes involved in a particular process inside the cell.
   d. help reveal the genetic basis of particular diseases.
   e. all of the above

10. Why would it be extremely difficult to develop genetic tests for all known human diseases?
    a. Only a very few diseases are known to have a genetic component.
    b. Many diseases involve mutations in more than one or two genes.
    c. It is difficult to obtain precise gene sequences.
    d. Current techniques are not powerful enough to detect mutations in single genes.
    e. all of the above

11. The Human Genome Project has uncovered the identity and location of over 10 million SNPs (single nucleotide polymorphisms). Through DNA chip technology, SNPs can be used to develop rapid genetic screening tests for a multitude of genetic disorders. However, not all SNPs would be useful for such a purpose because

a. SNPs cannot be used to detect gene mutations.
b. DNA chip technology can only detect a single gene at a time.
c. SNPs have not yet been associated with any human diseases.
d. only a small percentage of SNPs are found within gene coding sequences.
e. none of the above

12. The discovery that fruit flies and humans may share many genes involved in regulating the biological clock of the organism is an example of
a. functional genomics.
b. genetic screening.
c. behavioral genomics.
d. comparative genomics.
e. none of the above

Related Activities

- Perform a search on the World Wide Web for information about the Celera Genomics Corporation. Read through the overview and mission statement for the company to learn more about its vision for making the best use of the information obtained from the sequencing of the human genome. For comparison purposes, visit the U.S. Department of Energy's Human Genome Project Web site. Compose a one-page essay comparing and contrasting the differing viewpoints of these two organizations.
- Browse through your Sunday paper's health section. Count the number of articles related to human medicine. How many of these describe or reference the Human Genome Project? Summarize the articles you have found in terms of how each one describes the use of DNA sequence information either to investigate the cause of a disease or to develop a particular treatment.
- Using the World Wide Web, conduct a search for a list of currently available genetic tests. From this list, select 2–3 tests and discuss whether or not you would be willing to be tested to determine if you were a carrier for these diseases and the possibility for passing these disease alleles on to your children. Describe your reasons for being willing or not willing to be tested for these disorders.

Answers and Explanations

Factual Knowledge

1. **e** The first draft genome sequence was obtained from a bacterium (*H. influenzae*, 1995), followed thereafter by the budding yeast (1996), nematode (1998), fruit fly (2000), and human (2001). Refer to Figure C.2 for information on this subject.

2. **d** The field of genomics expands on the field of genetics, incorporating analysis of the entire genome and how it evolves over time.

3. **c** The Human Genome Project has revealed the identity and location of over 10 million SNPs, with more being discovered each day.

4. **e** 60 percent of human proteins are known to have fruit fly counterparts.

5. **d** The human genome contains roughly 3.2 billion base pairs. Refer to Table C.1 for a comparison with other organisms.

6. **False** Since the noncoding regions of DNA contain the regulatory elements for expression of the protein-coding genes, it is imperative to determine their sequences as well, in order to understand thoroughly how the genome operates as a whole.

7. **b** The Human Genome Project was comprised of a consortium of sequencing labs from around the world. Among the primary funders of the project were the National Institutes of Health and the U.S. Department of Energy.

8. **e** SNP profiling using DNA chip technology promises to provide rapid genetic screening capabilities. This will supply physicians with a powerful tool to assist them in developing a comprehensive health plan for their patients.

9. **False** The sequence reported in 2001 was merely a draft. Scientists are still working to fill in gaps in the sequence and validating the data collected.

10. **c** Based upon the completion of the human genome map and advances in DNA sequence analysis software, the revised estimate of the number of genes in the human genome is approximately 20,000 to 30,000, down from the previous estimate of 100,000.

11. **c** A protein domain is a segment that has a specific function, such as enzymatic activity. These conserved regions are useful for establishing the evolutionary relatedness of species.

12. **e** Genetic screening tests are available for at least one gene known to be linked to each of these disorders.

13. **d** single nucleotide polymorphism
a comparative genomics
b Human Genome Project
e DNA chip
c Celera Genomics

Conceptual Understanding

1. **c** Both genomics and genetics make use of DNA sequence analysis data. From a population and evolutionary genetics perspective, information obtained from DNA sequencing can provide insight into the genetic relatedness of organisms within a single population or in different populations. The field of genomics makes use of DNA sequences to develop clinically useful genetic screening tests and to assist in the discovery of new cures for genetically based diseases.

2. **e** DNA chip technology can be used to detect the presence of SNPs in a particular tissue, such as a cancer mass. This technology cannot, however, be used to determine if the tumor is malignant or benign or whether genes are expressed. DNA sequencing is the only technology that can be used to determine the sequence of DNA.

3. **d** Recall that there are currently hundreds of genetic tests available, including those for Huntington disease, cystic fibrosis, and breast cancer.

4. **b** The proposed second round of genome sequencing will target organisms such as the marmoset, the sea slug, the pea aphid, and several fungi. As such, these organisms would not be considered evolutionarily close to humans. These organisms potentially hold the key to understanding several genetic diseases and may help farmers grow better crops.

5. **a** To avoid the possibility of sequencing a single genome that may have contained a significant number of mutations, researchers working on the Human Genome Project used DNA samples from several anonymous individuals representing a cross section of the human population.

6. **e** All of the reasons listed were instrumental in the decision to begin the project with smaller, more manageable genome sequencing efforts. These efforts were successful in obtaining the complete sequences of organisms such as *E. coli*, *S. cerevisiae*, *C. elegans*, and *H. influenzae* and proved to be valuable in terms of the experience the scientists gained. This strategy facilitated the rapid completion of the first draft of the human genome years ahead of schedule.

7. **b** Recall that while behavioral genomics focuses on the genetic basis for the development of certain behavioral predispositions within a single group of organisms, the field of comparative genomics may also provide clues to such anomalies by comparing genomes across species.

8. **d** Since each gene can either be "on" or "off," the total number of combinations of gene products would be 2^N, where N = number of genes present in the organism. Therefore, Species A, with a total of 3 genes, would have $2^3 = 8$ possible RNA combinations, while Species B, with a total of 5 genes, would have $2^5 = 32$ possible RNA combinations: a fourfold increase in complexity. This example illustrates that while there was a recent revision in the estimate of the number of human genes present in the genome, the complexity of the human genome is immense (20–30,000). If one adds to this the variations produced through alternative RNA splicing and posttranslational modifications, it is easy to see how complicated it is to coordinate the development of an entire human being.

9. **e** DNA sequence information can provide scientists with a powerful tool to conduct direct comparisons with other species. In addition to yielding evolutionary data, such comparisons can provide insight into the function of genes whose role is unknown in humans but may be known in other species.

10. **b** Since many diseases involve mutations in more than a few genes, it is difficult to decipher and understand the connection between these gene defects and the causes of disease. This problem makes the development of reliable genetic tests difficult.

11. **d** Since only a small percentage of the human genome actually codes for protein, only a small proportion of identified SNPs would be located within these regions. As a result, the utility of the 10 million identified SNPs for genetic screening purposes would be limited.

12. **d** Recall that comparative genomics is the field that studies and compares genomes from multiple organisms in order to gain a better understanding of the function of various genes.

CHAPTER 16 How Evolution Works

Getting Started

Below are a few questions to consider prior to reading Chapter 16. These questions will help guide your exploration and assist you in identifying some of the key concepts presented in this chapter.

1. What was the name of the ship on which Charles Darwin sailed to the Galápagos islands?
2. What household item did the American Medical Association tell U.S. consumers to avoid using in 2000?
3. What do a person's arm, a whale's fin, and a bat's wing all have in common?
4. What can the fossil record tell us about how the Earth's environments have changed over time?
5. What are vestigial organs?
6. By what amount does the distance between South America and Africa increase each year?
7. How can computer programs that mimic natural selection help engineers to develop new inventions?

A Guide to the Reading

The following concepts typically give students the most difficulty when exploring the content in Chapter 16 for the first time. For each concept, one or more references have been identified that may help you gain a better understanding of these potentially problematic areas.

Mechanisms of Evolution

Evolution within a population occurs when individuals possessing certain characteristics survive and reproduce at a higher rate than other individuals. As described in the chapter, there are two primary means by which this occurs—natural selection and genetic drift. While both mechanisms result in the evolution of a population, they accomplish this in very distinct ways. In natural selection, the characteristics that an individual possesses actually contribute to the individual's survival and rate of reproduction. Since these characteristics may be inherited, subsequent generations would also possess these characteristics and also have higher survival and reproduction rates. Over several generations, this would cause the population to change to reflect the abundance of individuals with these favorable characteristics. In contrast, genetic drift occurs when natural events remove particular groups of individuals with particular characteristics by chance. As a result, surviving individuals with distinct characteristics go on to reproduce, passing their inheritable characteristics off to the next generation. The key to understanding the difference between these mechanisms is realizing that in genetic drift, inheritable characteristics play no role in determining survival. Survival of individuals with particular characteristics occurs solely by chance.

For more information on this concept, be sure to focus on:

- Section 16.2, Mechanisms of Evolution
- Figure 16.2, Natural Selection at Work
- Figure 16.3, Genetic Drift

Shared Characteristics

As discussed in the chapter, different species may share many characteristics. Features that are shared between organisms that have a common ancestor are called *homologous*. A good example would be the opposable thumbs shared by the great apes and humans. These two groups of organisms shared a recent common ancestor who likely also had this distinguishing feature. *Analogous* features, in contrast, appear when different groups of organisms develop similar characteristics separately through natural selection as a result of similar environmental pressures. This process is referred to as convergent evolution. The example provided in the text indicates that the features shared by the distantly related organisms sharks and dolphins (such as large dorsal fins) would have evolved through convergent evolution. As a result, these features would be considered analogous.

For more information on this concept, be sure to focus on:

- In Section 16.3, Organisms share characteristics due to common descent
- Figure 16.4, Shared Characteristics
- Figure 16.5, The Power of Natural Selection

Tying It All Together

Several concepts presented in this chapter build upon concepts presented in previous chapters and may also be revisited and discussed in greater detail in subsequent chapters, including:

Evolutionary Trees and Branching

- Chapter 2 – Section 2.2, Using Evolutionary Trees to Predict the Biology of Organisms

Genes and Alleles

- Chapter 10 – Section 10.3, Basic Patterns of Inheritance

Evolution of Populations

- Chapter 17 – Section 17.3, Four Mechanisms Can Cause Populations to Evolve

Mutation

- Chapter 17 – Section 17.4, Mutation: The Source of Genetic Variation

Genetic Drift

- Chapter 17 – Section 17.6, Genetic Drift: The Effects of Chance

Natural Selection

- Chapter 17 – Section 17.7, Natural Selection: The Effects of Advantageous Alleles

Using the Fossil Record to Determine Evolutionary Relationships

- Chapter 19 – Section 19.1, The Fossil Record: A Guide to the Past

Practice Questions

Factual Knowledge

1. Darwin's *The Origin of Species* first appeared in
   a. 1831.
   b. 1836.
   c. 1859.
   d. 1900.
   e. 1953.

2. According to the definition of *evolution*, which of the following has undergone this process?
   a. organisms
   b. hats
   c. cars
   d. computers
   e. all of the above

3. Biological evolution is best defined as
   a. the change in genetic characteristics of populations over time.
   b. random shifts in the genetic makeup of a population.
   c. changes in populations over time.
   d. characteristics that improve the chances of survival for a particular organism.
   e. allowing only individuals with certain characteristics to breed.

4. When natural selection is operating, which of the following is likely to occur?
   a. The characteristics of individuals will evolve over their lifetime.
   b. Individuals within a population may have greater survival than others.
   c. The evolution of a population is driven by random natural events.
   d. Mutations that occur in one cell are spread to the remainder of the cells in an organism.
   e. all of the above

5. Which of the following is an example of a shared characteristic between species that exists because of common evolutionary descent?
   a. the type of flower that bees will preferentially pollinate
   b. the number and type of bones in the limbs of vertebrates
   c. the similarity of color and pattern between frogs and their environment
   d. mealybugs' ability or inability to feed on genetically engineered corn
   e. all of the above

6. Modern evidence strongly supports the evolutionary prediction that organisms should possess signs of their common ancestry with other organisms. (True or False)

7. With which of the following statements would Darwin most likely *disagree*?
   a. Individuals within a population vary in the characteristics they possess.
   b. Evolution is best viewed as a purposeful and directed change over time.
   c. Natural selection is the mechanism by which biological evolution takes place.
   d. The fossil record supports the view that biological evolution has occurred.
   e. all of the above

8. Science has never actually observed the process of evolutionary change in species. (True or False)

9. Most religious leaders today view evolution and religion as compatible. (True or False)

10. The evidence supporting biological evolution includes
    a. fossils.
    b. patterns of embryological development.
    c. shared anatomical characteristics.
    d. molecular similarities.
    e. all of the above

11. Which of the following species is a product of artificial selection?
    a. horse
    b. tuna
    c. broccoli
    d. medium ground finch
    e. corn mealybug

12. Refer to Figure 16.12 in the textbook. Which of the following do all of these birds have in common?
    a. They live in Africa.
    b. Darwin knew nothing about them.
    c. They are the product of artificial selection.
    d. All are descended from a common ancestor.
    e. Their origins can be explained by continental drift.

13. Which of the following does *not* have a direct effect on the process of biological evolution?
    a. techniques for determining the age of fossils
    b. differences in reproductive success
    c. variation in the inheritance of traits
    d. changes in the environment
    e. natural selection

14. Match each term with the best description.
    __ adaptation
    __ evolution
    __ natural selection
    __ genetic drift
    __ speciation
    a. change through time
    b. random change in the genetics of a population
    c. splitting of one type of organism into others
    d. mechanism that drives biological evolution
    e. beneficial characteristic of an organism

Conceptual Understanding

1. Many bizarre plants and animals inhabit the Galápagos Islands, as young Charles Darwin observed when he visited there in the early 1830s. Which of the following is the most likely reason that oceanic islands often have such unusual inhabitants?
    a. Natural selection does not work on islands.
    b. Evolution takes place more slowly on islands than it does elsewhere.
    c. Islands are isolated and thus susceptible to strong natural selection.
    d. Islands usually have many more species than do mainland habitats.
    e. Artificial selection is more common on islands than on the mainland.

2. Among individuals, the ultimate source of heritable differences fueling natural selection is
    a. gene mutation.
    b. industrial pollution.
    c. continental drift.
    d. all of the above
    e. none of the above

3. The evolutionary effects of genetic drift are
    a. most obvious in large populations.
    b. usually random and unpredictable.
    c. always the same as those of natural selection.
    d. shaped by nonrandom selective forces.
    e. exemplified by population changes in the peppered moth.

4. With which of the following statements would an evolutionary biologist *agree*?
    a. The fossil record is the only source of evidence in favor of evolution.
    b. Evolving populations rarely show changes in their genetic makeup.
    c. There has been no significant change in organisms through time.
    d. Mutations provide the foundation for evolutionary change.
    e. God was the designer of all organisms in the form in which they now appear.

5. Anteaters and some whales lack teeth as adults. The presence of teeth in the embryos of these creatures is evidence that
    a. all animals must undergo some type of embryological development.
    b. the embryos have a very different diet from that of the adults.
    c. these species are the product of artificial selection.
    d. these species share a common ancestry with animals that have adult teeth.
    e. none of the above

6. The cytochrome c data in Figure 16.7 in the textbook suggests that which of the following statements is *true*?
    a. Rhesus monkeys and humans are fairly closely related.
    b. Humans and moths are only distantly related.
    c. Tuna and dogs are only distantly related.
    d. all of the above
    e. none of the above

7. Again refer to Figure 16.7 in the textbook. If you were to plot the cytochrome c data for a frog on this graph, where would it most likely be placed in relationship to the other animals?
    a. between the human and the rhesus monkey
    b. between the rattlesnake and the dog
    c. between the tuna and the rattlesnake
    d. between the moth and the tuna
    e. to the left of the moth

8. Based on the information in Figure 16.4a in the textbook, which of the following organisms should have five digits and the same set of arm bones?
   a. chimpanzee
   b. cat
   c. squirrel
   d. dolphin
   e. all of the above

9. The evolution of horses shown in Figure 16.6 in the textbook is a good example of the evolutionary process because it
   a. provides solid fossil evidence in favor of evolution.
   b. lacks a number of key specimens and so is rather fragmented.
   c. shows that Earth is indeed only 10,000 years old.
   d. demonstrates a decrease in body size between *Hyracotherium* and *Equus*.
   e. none of the above

10. Consider two different species of turtle. These organisms
   a. are members of the same population.
   b. are capable of breeding with each other.
   c. share very few anatomical characteristics.
   d. were likely formed through natural selection.
   e. none of the above

11. The rudimentary hind legs of a python and the fully functional legs of other reptiles are
   a. shared characteristics that result from common descent.
   b. freaks of development that have nothing to do with evolution.
   c. the result of developmental abnormalities that sometimes occur.
   d. examples of fossil evidence in favor of evolution.
   e. not the products of natural selection.

12. The Galápagos Islands have played a significant role in the development of evolutionary thinking for more than 150 years. Which of the following is not an example of evolutionary change that has been studied using species from the Galápagos?
   a. short-term species adaptation to drought conditions
   b. breeding of new plant varieties from a single wild species
   c. results that occur when a new species finds unused habitat
   d. effects of geographic isolation on the process of natural selection
   e. the effects of environmental pressures on the adaptations of individual species

Related Activities

- Use your school library or the World Wide Web to research Charles Darwin and his ideas. Compose a one-page essay about his educational background and how his scientific ideas were developed. Be sure to consider the historical context of society at the time Darwin introduced his ideas about evolution and how that societal perspective influenced the initial acceptance of *The Origin of Species*.
- Using the World Wide Web, find an example other than horses that shows good fossil evidence for evolution. Compose a one-page summary explaining the example you found.
- Review the Biology Matters section "You Can't Live with 'Em, You Can't Live without 'Em," which deals with the overuse of antibacterial products. Perform a search on the World Wide Web for information on the evolution of antibiotic resistance and the problem this has caused for the medical community. Compose a one-page summary of your results focusing on the role natural selection plays in this evolutionary phenomenon.

Answers and Explanations

Factual Knowledge

1. **c** Darwin set sail on the *Beagle* in 1831 and returned to England in 1836. He published *The Origin of Species* in 1859.
2. **e** Evolution is simply descent with modification, and all of these things have done that. Only organisms, however, have undergone biological evolution.
3. **a** The idea of genetic change over time is key to understanding biological evolution.
4. **b** One of the keys to natural selection is the enhanced survival of individuals with particular inheritable characteristics that can be passed on to subsequent generations.
5. **b** The simplest explanation is that this limb bone pattern is an ancestral characteristic that evolved at the beginning of the vertebrate lineage, and thus formed the basis for the body shape of all groups that followed.
6. **True** Examples include internal anatomy, details of molecular structure, and embryological patterns.

7. **b** There is nothing purposeful or directed about natural selection and its evolutionary outcomes. There is only a filtering of possible adaptations using the backdrop of environmental pressures that change on a regular basis.

8. **False** Scientists observe change in the genetic makeup of populations (that is, biological evolution) all the time. There are also examples of new plant and animal species that have appeared within the history of modern science.

9. **True** As indicated in the chapter, the Catholic Church has accepted that evolution may explain the physical characteristics of humans, but that religion is necessary to explain our spiritual characteristics.

10. **e** All of these areas contribute evidence in support of biological evolution.

11. **c** Broccoli is descended from a species of wild mustard that has also given rise under domestication to other edible plants such as cauliflower and cabbage. All of these plants are the same species. Refer to Figure 16.9 for additional information.

12. **d** The Galápagos finches have evolved into a variety of forms, but all are descended from a single species that came from the mainland some time in the evolutionary past.

13. **a** While these techniques give us a better picture of the way that fossil organisms are related to each other chronologically, they are not themselves a part of the evolutionary process.

14. **e** adaptation
a evolution
d natural selection
b genetic drift
c speciation

Conceptual Understanding

1. **c** Because they are isolated, islands often have very specific selective pressures that can dramatically shape adaptations.

2. **a** Gene mutation is the only way that new alleles can form, and new alleles are what lead to new, and potentially adaptive, variations.

3. **b** Genetic drift is the random and unpredictable change in the genetic makeup of a population due to small population sizes or chance events.

4. **d** Heritable differences are required for evolution to occur. These heritable differences are generated through gene mutation.

5. **d** The loss of teeth after embryological development suggests that these species are descended from toothed ancestors, and that they have subsequently lost their adult teeth as an adaptive response to selective pressures in their environment.

6. **d** All of these are correct interpretations of the data based on the degree of similarity in cytochrome *c* structure indicated. Species with the fewest mutations between them are closely related, while those with large differences are more distantly related.

7. **c** Frogs are amphibians and are intermediate between fish and reptiles in terms of evolutionary origins. Their cytochrome *c* structure reflects this evolutionary history.

8. **e** Since all are mammals, they too should show the same basic pattern of limb bones seen in humans, whales, and bats.

9. **a** Only some of the fossil evidence now available for horses is shown in Figure 16.6. Even without these additional data, it is clear how horses represent a steady progression from smaller to larger animals.

10. **d** As best we know, natural selection is the mechanism by which virtually all new species arise.

11. **a** Since both of these species share some form of the feature, it must have been present in a common ancestor.

12. **b** This would be similar to the example of wild mustard plants that were artificially bred to produce cabbage and other related varieties. Such studies have not been part of the Galápagos research history.

CHAPTER 17 Evolution of Populations

Getting Started

Below are a few questions to consider prior to reading Chapter 17. These questions will help guide your exploration and assist you in identifying some of the key concepts presented in this chapter.

1. What caused the U.S. Surgeon General to declare that it was time to "close the book on infectious disease" in 1960?
2. How many species of insects are estimated to be resistant to at least one pesticide?
3. What causes human immunodeficiency virus (HIV) to act like a "moving target" in an infected individual's body?
4. What do the Florida panther, northern elephant seal, and African cheetah all have in common?
5. What legislation has caused the directional selection of light-colored moths over dark-colored moths in England over the past several decades?
6. What is the ideal birth weight for human babies as determined by stabilizing selection?

A Guide to the Reading

The following concepts typically give students the most difficulty when exploring the content in Chapter 17 for the first time. For each concept, one or more references have been identified that may help you gain a better understanding of these potentially problematic areas.

Genetic Drift

Genetic drift is the process by which alleles are selected within a population at random over time. The frequency of particular alleles may change as a result of random events. This may have a profound effect on the population if the size of the population happens to be small. Since small populations have fewer total alleles, if a number of individuals happen to leave or are eliminated from the population, the relative effect on the remaining alleles can be drastic. When one allele reaches a frequency of 100 percent as a result of the genetic drift of other alleles, the remaining allele is said to have reached "fixation," that is, because this is the only allele present in the population, it will not change over time. It is important to note that the process of genetic drift is not subject to adaptive evolution; it may result in the fixation of alleles that are beneficial, neutral, or even harmful to the population.

For more information on this concept, be sure to focus on:

- In Section 17.6, Genetic drift affects small populations
- Figure 17.3, Genetic Drift

Types of Natural Selection

As described in the chapter, there are three different types of natural selection, each serving to shape the evolution of a population as a result of differences in the survival and reproductive success of members possessing particular traits. Directional selection occurs when individuals possessing an extreme trait have a reproductive or survival advantage over individuals lacking this trait. An example of directional selection might include the neck length of giraffes. Giraffes with longer necks might have a survival advantage over animals with shorter necks since they may be more adept at feeding from tall trees. The characteristic determining neck length would then be inherited by subsequent generations, shifting the characteristic of the population to possess longer necks. Stabilizing selection occurs when individuals with an intermediate trait have a survival or reproductive advantage over individuals with extreme versions of the trait. The example provided in the chapter focuses on human birth weight; underweight and/or premature babies have a survival disadvantage while very large babies are likely to cause problems during delivery. These factors have stabilized the average birth weight of babies at around 8 pounds. Disruptive selection occurs when individuals possessing a trait at either extreme have a survival or reproductive advantage over individuals with intermediate forms of the trait. The example provided in the chapter focuses on the beak size within a population of African seed cracker birds. The advantage of having either a large or a small beak is that it enhances feeding efficiency, allowing populations to feed on either hard or soft seeds. Another way to look at this is to realize that individuals possessing intermediate-sized beaks would be at a disadvantage since they would be inefficient at feeding on both types of seeds.

For more information on this concept, be sure to focus on:

- In Section 17.7, There are three types of natural selection
- Figure 17.6, The Three Types of Natural Selection
- Figure 17.7, Directional Selection in the Peppered Moth
- Figure 17.8, Stabilizing Selection for Human Birth Weight
- Figure 17.9, Disruptive Selection for Beak Size

Tying It All Together

Several concepts presented in this chapter build upon concepts presented in previous chapters and may also be revisited and discussed in greater detail in subsequent chapters, including:

Genetic Recombination

- Chapter 11 – Section 11.3, Linkage and Crossing-Over

Mutations

- Chapter 12 – Section 12.4, Repairing Replication Errors and Damaged DNA

Microevolution

- Chapter 16 – Section 16.1, Biological Evolution: The Sum of Genetic Changes

Genetic Drift and Natural Selection

- Chapter 16 – Section 16.2, Mechanisms of Evolution

Adaptation

- Chapter 18 – Section 18.1, Adaptation: Adjusting to Environmental Challenges

Speciation

- Chapter 18 – Section 18.4, Speciation: Generating Biodiversity

Practice Questions

Factual Knowledge

1. Which of the following is an example of rapid evolutionary change in a population?
   a. genetic bottlenecks in prairie chickens
   b. sexual selection in gophers and pine trees
   c. pesticide resistance in insects
   d. average birth weight in humans
   e. all of the above

2. Microevolution refers to
   a. the evolution of micoorganisms.
   b. evolution that occurs on a large scale.
   c. the changes in allele or genotype frequencies that occur in a population over time.
   d. the splitting of a large group of organisms into multiple species.
   e. none of the above

3. If there are 1000 individuals in a population and 600 of them are homozygous dominant for a gene that has only two alleles, what is the genotype frequency for the homozygous dominant condition?
   a. 1.0
   b. 0.6
   c. 0.5
   d. 0.4
   e. 0.3

4. A particular gene in a population has two alleles, C and c. If the allele frequency of $C = 0.7$, what is the frequency of c?
   a. 1.0
   b. 0.7
   c. 0.3
   d. 0.0
   e. cannot determine from the information given

5. Which of the following statements about genetic variation is *true*?
   a. Mutation is the ultimate source of new alleles.
   b. The only source of genetic variation is mutation.
   c. New alleles are formed during sexual reproduction.
   d. Mutations that result in the production of new alleles are quite common.
   e. all of the above

6. The Hardy–Weinberg equation describes the genotype frequencies that will occur in an evolving population of organisms. (True or False)

7. In the Hardy–Weinberg equation, the terms p and q represent the ________ in a population.
   a. size of the gene pool
   b. total number of alleles
   c. genotype frequencies
   d. allele frequencies
   e. mutation rates

8. A violation of which of the following conditions would cause a population to shift away from its stable Hardy–Weinberg frequencies?
   a. no genetic drift
   b. no gene flow
   c. no net mutation
   d. no natural selection
   e. any of the above

9. Sexual selection in a population
   a. is a special type of natural selection.
   b. always favors individuals with a higher survival potential.
   c. alters allele frequencies in a stable population.
   d. favors individuals with poor reproductive potential.
   e. reduces genetic variation in the population.

10. Which of the following statements about gene flow is *true*?
    a. It reduces the genetic variation within populations.
    b. It reduces the genetic variation between populations.
    c. It reduces the movement of alleles between populations.
    d. It does not violate the conditions of the Hardy–Weinberg model.
    e. It rarely occurs in natural populations.

11. Which of the following is more likely to occur when a population of organisms is small?
    a. natural selection
    b. mutation
    c. heterozygote advantage
    d. genetic drift
    e. random mating

12. Of the organisms listed below, which one is *not* believed to have undergone a genetic bottleneck at some point in its recent evolutionary history?
    a. *Culex* mosquitoes
    b. Northern elephant seals
    c. Illinois prairie chickens
    d. African cheetahs
    e. Florida panthers

13. DDT resistance in mosquitoes is one dramatic example of the mode of natural selection known as
    a. disruptive selection.
    b. directional selection.
    c. stabilizing selection.
    d. heterozygote advantage.
    e. genetic drift.

14. Match each term with the best description.
    __ allele frequency
    __ gene flow
    __ genotype frequency
    __ microevolution
    __ stabilizing selection
    a. p^2 in the Hardy–Weinberg equation
    b. short-term changes in allele frequencies within a population
    c. shifts toward the mean in a population
    d. proportion of a particular allele in the gene pool
    e. occurs as a result of migration between populations

Conceptual Understanding

1. The genetic mutations that occur in organisms are
   a. relatively rare events that have little consequence for evolution.
   b. predictable by scientists before they actually occur.
   c. directed toward a particular adaptive goal.
   d. the raw material of evolution.
   e. none of the above

2. The impact that a mutation has on a population depends largely on the environment in which the mutation occurs. (True or False)

3. Consider a population that has reached its Hardy–Weinberg equilibrium frequencies. If this population is suddenly subjected to strong directional selection, both the allele and the genotype frequencies are likely to undergo change as a result. (True or False)

4. Consider a population with a total of 500 alleles for a specific gene. How many individuals are in this population?
   a. 1000
   b. 500
   c. 250
   d. 125
   e. cannot determine from the information given

5. A population of organisms has a gene for which there are two alleles, *D* and *d*. The allele frequency of $D = 0.8$. If this population satisfies all five of the Hardy–Weinberg conditions, what are the genotype frequencies that are expected in the next generation?
   a. $DD = 0.04$; $Dd = 0.32$; $dd = 0.64$
   b. $DD = 0.64$; $Dd = 0.32$; $dd = 0.64$
   c. $DD = 0.04$; $Dd = 0.64$; $dd = 0.32$
   d. $DD = 0.64$; $Dd = 0.32$; $dd = 0.04$
   e. cannot determine from the information given

6. Several generations into the future, a population of birds currently undergoing stabilizing selection will look
   a. just like it does now.
   b. more like the current mean.
   c. more like one of the extremes.
   d. more like both of the extremes.
   e. none of the above

7. In the Hardy–Weinberg equation, the term $2pq$ represents the
   a. overall gene frequency of the population.
   b. frequency of both homozygous genotypes.
   c. frequency of the heterozygous genotype.
   d. allele frequencies of the population.
   e. effect of disruptive selection on the population.

8. Refer to Figure 17.3 in your textbook. If the population of wildflowers in this hypothetical simulation of genetic drift were made up of 1000

individuals instead of 10, how would this affect the change in allele frequencies?
   a. There would probably be a much smaller change in allele frequencies.
   b. There would definitely be a much greater change in allele frequencies.
   c. There would be no difference between the two types of simulation.
   d. The *aa* genotype would come to dominate by generation III.
   e. The *Aa* genotype would come to dominate by generation III.

9. A population susceptible to genetic bottlenecks would have which of the following characteristics?
   a. small population size
   b. high mutation rates
   c. high levels of gene flow
   d. sexual selection
   e. evolution by natural selection

10. Which of the following would *not* be considered as one of the steps in the evolutionary process?
    a. Random mutations and genetic rearrangements occur.
    b. Natural selection acts upon the genetic variation present in a population.
    c. Inheritable characteristics are produced by random genetic events such as mutation.
    d. Allele frequencies in a population change over time.
    e. none of the above

11. In Figure 17.8 in your textbook, the survival of newborn humans is compared to their birth weight. Note that these data are from London and span the time period between 1935 and 1946. If you were to look at a graph of comparable data for the same population from the 1990s, what differences, if any, would you see?
    a. There would be no differences.
    b. There would be a broader optimum birth weight due to medical advances.
    c. There would be a sharper optimum peak due to medical advances.
    d. The range of weights over which newborns survive would have decreased.
    e. The length of time that underweight newborns survive would have decreased.

12. Which of the following contributes to the spread of antibiotic resistance among pathogenic bacteria?
    a. the transfer of resistance genes between bacteria
    b. poor sanitation in areas where pathogenic microbes are plentiful
    c. indiscriminate use of antibiotics
    d. directional selection
    e. all of the above

Related Activities

- Search your library or the World Wide Web to learn more about the problem of antibiotic resistance, which is described in the chapter. Compose a one-page summary of the results you have identified and propose a course of action our society should take to help control this growing problem. Be sure to focus on the economic and political factors that must be taken into account when addressing this problem.
- In the battle over evolution versus creationism, one of the biggest problems is the comparison of microevolution with macroevolution. Your textbook has provided abundant evidence that microevolution occurs, as most creationists would acknowledge. However, creationists object to the idea that macroevolution naturally follows from microevolutionary change. What is your view about this debate? Is it reasonable to think that significant changes in species type can emerge from the accumulated small changes in allele frequencies that occur between populations? You may want to search the World Wide Web at both pro-creation and pro-evolution sites to gather more information when developing your opinion.
- Consider the example provided in the chapter for directional selection in the Peppered Moth. Using the World Wide Web, identify at least one other example of directional selection that has been influenced by the activities of humankind. With your example, compose a one-page essay indicating how humankind has influenced the evolution of the species cited and what steps could be taken to restore the population to its original composition.

Answers and Explanations

Factual Knowledge

1. **c** The gene for pesticide resistance in mosquitoes is thought to have spread throughout major populations of these insects in only a few decades. All the other answers, while related to evolution, are not examples of rapid evolutionary change.

2. **c** Microevolution refers to the changes in allele or genotype frequencies that represent the smallest scale at which evolution can occur.

3. **b** The genotype frequency is simply the proportion of that genotype in the population, in this case 600/1000 =

0.6. Since the genotype frequencies must add up to 1.0 (all genotypes), the remaining 0.4 is made up of the heterozygotes and homozygous recessive individuals.

4. **c** Just like genotype frequencies, allele frequencies must add up to equal 1.0 (all alleles). Therefore, 1.0 - 0.7 = 0.3.
5. **a** While mutation is the only source of new alleles, it is not the only source of genetic variation. Sexual recombination also produces variation, but it does not cause mutation. Finally, mutation rates are generally low in most populations.
6. **False** The Hardy–Weinberg equation describes the situation of unchanging genotype and allele frequencies and is valid only when a population is *not* evolving.
7. **d** These allele frequencies can be used in the Hardy–Weinberg equation to generate expected genotype frequencies for the next generation of a nonevolving population.
8. **e** All of these conditions are necessary for the validity of the predictions generated by the Hardy–Weinberg equation.
9. **a** Recall that sexual selection occurs when a characteristic or set of characteristics makes individuals better at finding mates. As a result, these characteristics may be more readily inherited by subsequent generations. However, some of the characteristics that allow for better mating potential may actually be detrimental.
10. **b** Gene flow moves alleles between populations, thus making those populations more alike genetically.
11. **d** Genetic drift is the random change in allele frequencies that occurs when a population is small. It stems from the strong influence of chance events when sample sizes are small.
12. **a** Although the data for panthers, cheetahs, and elephant seals are not as clear-cut as the data for prairie chickens, these animals probably did suffer relatively recent genetic bottlenecks. Mosquitoes are interesting for other population-level evolutionary events.
13. **b** See Figures 17.6 and 17.7 in your textbook for information on directional selection.
14. **d** allele frequency
    e gene flow
    a genotype frequency
    b microevolution
    c stabilizing selection

Conceptual Understanding

1. **d** Although relatively rare, mutations are critical to evolution because they are the ultimate source of new alleles. They are not directed, nor are they predictable.
2. **a** Whether a mutation is deleterious, neutral, or beneficial depends largely on the way the environment in which an organism lives "interprets" the phenotypic effect of the mutation.
3. **a** Directional selection favors one extreme of a population. This will favor one genotype over others and likely result in a shift in both genotype and allele frequencies.
4. **c** Since each diploid individual has two alleles, the number of individuals is one-half the total number of alleles in the population.
5. **d** We are told that the allele frequency for $D = 0.8$; therefore, the frequency of $d = 0.2$. These values are the p and q that we need to calculate the genotype frequencies in the next generation. Using the Hardy–Weinberg equation, p^2 (*DD*) = 0.64, $2pq$ (*Dd*) = 0.32, and q^2 (*dd*) = 0.04.
6. **b** Stabilizing selection shifts the population extremes toward the mean.
7. **c** The terms p and q are the allele frequencies for a given gene, so $2pq$ represents the two possible ways that heterozygotes can form.
8. **a** Genetic drift is a product of chance events that affect small populations much more than large populations. Therefore, in a population of 1000 wildflowers, we would likely not see much change in allele frequencies due to genetic drift.
9. **a** As with genetic drift, bottlenecks are a product of small population sizes.
10. **e** All of the events listed can be considered as steps in the evolutionary process.
11. **b** Medical advances now help underweight babies survive and allow large babies to be delivered surgically. Thus the stabilizing selection of natural childbirth has been lessened, which makes the curve less peaked.
12. **e** The emergence of antibiotic-resistant strains of bacteria is due to the strong directional selection forces that act when antibiotics are used indiscriminately. The genes that confer resistance have been shown to spread among populations of bacteria.

CHAPTER 18 Adaptation and Speciation

Getting Started

Below are a few questions to consider prior to reading Chapter 18. These questions will help guide your exploration and assist you in identifying some of the key concepts presented in this chapter.

1. What has caused the disappearance of over 200 Lake Victoria cichlid species over the last 30 years?
2. How is the practice of dog breeding similar to the process of adaptive evolution?
3. What unique adaptations of the "four-eyed fish" have enabled this species to survive in its environments?
4. In terms of speciation, how does reproductive isolation differ from geographic isolation?
5. What is a ring species?
6. Which animal species have been known to develop as a result of polyploidy?
7. What role does adaptation play in our fight against HIV infection?

A Guide to the Reading

The following concepts typically give students the most difficulty when exploring the content in Chapter 18 for the first time. For each concept, one or more references have been identified that may help you gain a better understanding of these potentially problematic areas.

Reproductive Isolation

As defined in the chapter, a species is a group of organisms that are capable of interbreeding with each other. Different species remain as separate groups because they are not able to reproduce with members of other species. This can be because of either geographical isolation or reproductive isolation. Individuals from different species are incapable of interbreeding; therefore they do not exchange genetic information, and they remain genotypically and phenotypically distinct. Species are said to be *reproductively isolated* from other species as a result of these reproductive barriers. Reproductive barriers can take several forms, including those that prevent successful mating (due to ecological, behavioral, mechanical, or gametic isolation) and those that prevent the survival or the success of hybrid species (zygote death, hybrid performance) when mating between two species occurs.

For more information on this concept, be sure to focus on:

- In Section 18.3, Species are reproductively isolated from one another
- Table 18.1, Barriers That Can Reproductively Isolate Two Species in the Same Geographic Region

Allopatric vs. Sympatric Speciation

Speciation occurs when a single species evolves into two or more distinct species over a period of time. The two main mechanisms by which speciation occurs are allopatric speciation and sympatric speciation. The key to understanding the difference between these mechanisms is to realize that allopatric speciation occurs when new species are formed as the result of the geographic isolation of populations of the original species. Over time, these separated populations may evolve independently and become distinct in terms of both genotype and phenotype. It is important to note that the distance separating the two populations must be great enough to inhibit gene flow between the populations. This factor makes the distance needed to achieve geographic separation vary depending upon the species (e.g., organisms capable of flight will be able to easily cross new rivers or canyons). Sympatric speciation is the most common mechanism by which species form. It is important to note that sympatric speciation does not involve the geographic isolation of groups. Instead, other factors, such as polyploidy (multiple sets of chromosomes) or the natural selection of alleles that favor the use of one food source over another, may result in the formation of distinct species within the same geographic region.

For more information on this concept, be sure to focus on:

- In Section 18.4, Speciation often results from geographic isolation
- In Section 18.4, Speciation can occur without geographic isolation
- Figure 18.9, Allopatric Speciation

Tying It All Together

Several concepts presented in this chapter build upon concepts presented in previous chapters and may also be revisited and discussed in greater detail in subsequent chapters, including:

Meiosis and Chromosome Number (Ploidy)

- Chapter 9 – Section 9.4, Meiosis: Halving the Chromosome Number

Genetic Control of Development

- Chapter 14 – in Section 14.3, Development in eukaryotes relies on gene cascades and homeotic genes

Evolution

- Chapter 16 – Section 16.1, Biological Evolution: The Sum of Genetic Changes

Gene Flow

- Chapter 17 – Section 17.5, Gene Flow: Exchanging Alleles between Populations

Natural Selection

- Chapter 17 – Section 17.7, Natural Selection: The Effects of Advantageous Alleles

Practice Questions

Factual Knowledge

1. Your textbook's discussion of the many species of cichlids living in Lake Victoria illustrates that
   a. natural selection can improve the match between organism and environment.
   b. one sex can drive reproductive isolation of species.
   c. the environment can influence the success of a particular species.
   d. natural selection can operate quite rapidly.
   e. all of the above

2. The selective pressures that have served to increase the number of cichlid species in Lake Victoria are
   a. predation and feeding.
   b. predation and mate attraction.
   c. mate attraction and feeding.
   d. feeding and making nests.
   e. all of the above

3. Which of the following statements about adaptations is *false*?
   a. Adaptations provide a close match between the organism and its environment.
   b. Many adaptations appear complex, even though they are caused by simple mechanisms.
   c. All adaptations help the organism accomplish important functions.
   d. Examples of adaptations are difficult to find in the natural world.
   e. Some adaptations are associated with improved reproduction.

4. The degree of genetic variation present in a species is one of the factors that potentially limits how much the quality of an adaptation can be improved. (True or False)

5. Developmental limitations on adaptation are due in part to
   a. geographic and reproductive isolation.
   b. the unintended effects of adaptive evolution.
   c. the many different effects of relatively few developmental genes.
   d. whether or not the individual reproduced during the previous year.
   e. all of the above

6. Because adaptations are often compromises between conflicting ecological pressures, most adaptations will likely be
   a. poor matches with the environment.
   b. good matches with the environment, although not perfect.
   c. excellent matches with the environment and nearly perfect.
   d. absolutely perfect matches with the environment.
   e. unaffected by adaptive evolution.

7. Figure 18.5 in your textbook provides an example of
   a. conflicting selective pressures that limit adaptation.
   b. the similarities in adaptation within a population of organisms.
   c. geographic isolation and the formation of hybrids.
   d. reproductive isolation due to a lack of gene flow.
   e. all of the above

8. According to your textbook's definition of the term *species*, the two types of oak pictured in Figure 18.7 are members of the same species. (True or False)

9. Which of the following is *not* a barrier that can lead to reproductive isolation between species living in the same geographic region?
   a. behavioral isolation
   b. zygote death
   c. hybrid performance
   d. hybridization
   e. ecological isolation

10. A definition of the term *species* based on reproduction has distinct limitations, which include its being
    a. of little use when defining fossil species.
    b. inapplicable to species that use asexual reproduction.
    c. of little use when organisms produce hybrids.
    d. unclear and imprecise.
    e. all of the above

11. Which of the following *cannot* occur when a population is in reproductive isolation from others of its kind?
    a. gene flow
    b. mutation
    c. natural selection
    d. nonrandom mating
    e. genetic drift

12. Which of the following groups demonstrates the formation of ring species as a result of geographic isolation?
    a. birds of New Guinea
    b. salamanders of California
    c. plants in Hawaii
    d. cichlids in Africa
    e. all of the above

13. Speciation in animals can occur in the absence of geographic isolation. (True or False)

14. Match each term with the best description.
    __ allopatric speciation
    __ domestication
    __ polyploidy
    __ speciation
    __ hybrid
    __ sympatric speciation
    a. multiple sets of chromosomes
    b. results from a cross between related species
    c. a major cause of living diversity
    d. occurs without geographic isolation
    e. conversion of a wild species to human control
    f. reproductive isolation is needed to get this

Conceptual Understanding

1. As species, what do the weaver ant and moth caterpillar examples discussed in your textbook have in common?
    a. The behavior and morphology of both are the result of natural selection.
    b. They are both the result of recent speciation events.
    c. Neither species existed prior to the 1950s.
    d. Both are examples of domesticated species.
    e. They are both adapted for life on coastal sand dunes.

2. If a murky lake is stocked with a number of species of cichlid fish from Lake Victoria, what is most likely to happen?
    a. The cichlid species will be very visible to the many predators in this area.
    b. The male cichlid will be very attractive to females in the lake.
    c. The cichlid species will interbreed because the lake is so cloudy.
    d. The cichlid species will have many offspring because they are so colorful.
    e. none of the above

3. If a guppy species found in a stream is placed in a lake, its color will evolve rapidly from colorful to drab over as few as 10 generations because of the
    a. lack of food.
    b. presence of additional predators in the lake.
    c. temperature of the air.
    d. taste of the water.
    e. none of the above

4. Consider the caterpillars pictured in Figure 18.1 of your textbook. These two organisms are
    a. members of the same species.
    b. matched well with their environment.
    c. influenced developmentally by chemicals in their food.
    d. the larval stage of a moth.
    e. all of the above

5. Reproductive isolation can occur in a number of ways, but the net effect is always the same in that
    a. hybridization must take place before new species can form.
    b. it causes geographic isolation between organisms.
    c. the rate of speciation slows down.
    d. few or no genes flow between populations.
    e. females survive longer than males.

6. When different species mate and produce fertile offspring, those offspring
    a. are automatically considered new species.
    b. are said to be the product of hybridization.
    c. usually are more successful than their parents.
    d. show the characteristics of only one parent.
    e. all of the above

7. Which of the following must take place for speciation to occur?
    a. hybridization
    b. geographic isolation
    c. polyploidy
    d. reproductive isolation
    e. all of the above

8. Geographic isolation is said to occur whenever populations are separated by a distance that is great enough to limit ________.
    a. hybridization
    b. gene flow
    c. polyploidy
    d. domestication
    e. none of the above

9. Reproductive isolation is to speciation as natural selection is to
    a. adaptive evolution.
    b. hybridization.
    c. ecological isolation.
    d. polyploidy.
    e. building an ant nest.

10. Polyploidy is believed to be in some way responsible for the origin of over half of the animal species alive today. (True or False)

11. According to the definition of *species* used in your textbook, all of the dogs shown in the Biology on the Job box in this chapter are
    a. the result of polyploidy.
    b. geographically isolated.
    c. the same species.
    d. hybrids from the mating of two wolves.
    e. perfectly matched to their environment.

12. If you were to randomly select two species that have a common ancestor and measure the time that it took for them to diverge from the ancestor, your result would likely be
    a. months.
    b. a few years.
    c. hundreds of years.
    d. a few thousand years.
    e. millions of years.

13. Although populations of North American apple maggot flies are not geographically isolated, they appear to be in the process of diverging into two new species because the two populations
    a. feed on two different types of food.
    b. reproduce at different times.
    c. lay their eggs on specific food types.
    d. have very little gene flow between them.
    e. all of the above

14. The distance needed for geographic isolation is the same for a crow as it is for a snail. (True or False)

15. What do the fight against disease and the development of new crop species have in common?
    a. They both involve the hybridization of species.
    b. They both involve the adaptive evolution of species.
    c. They both require geographic isolation between species.
    d. Neither involves reproductive isolation between species.
    e. Neither involves the occurrence of polyploidy.

Related Activities

- Refer to the Biology in the News story about the Hobbit-sized humans discovered in Indonesia. Consider how geographic and/or reproductive isolation likely contributed to the formation of this human species. Compose a one-page analysis of how you believe *Homo sapiens* and *Homo floresiensis* might have co-existed today providing the latter species was still present. Assuming the two species were able to produce hybrids, what other factors might have contributed to the reproductive isolation of the two groups?
- Using your library or the World Wide Web, identify an example of animal domestication other than those described in the textbook. Compose a one-page summary discussing how and when the domestication of this animal took place and what changes in the species have resulted from human-controlled breeding programs.
- Search the World Wide Web for an example of a species that scientists believe may currently be in the process of speciation. In a one-page essay, describe the biology of this species, and discuss why scientists think that it may be about to give rise to multiple species.

Answers and Explanations

Factual Knowledge

1. **e** The effect of male cichlid body coloration and female mate choice on reproduction along with the difference in jaw structure among the immense number of species and the presence of predators and pollution in Lake Victoria illustrate all of these points.
2. **c** The tremendous number of species evolved because of mate selection and the evolution to specialize on various food items. Recall that predation by the Nile perch serves to reduce the number of species.
3. **d** Adaptations that match the organism to its environment are everywhere in nature.
4. **True** If adaptive evolution is going to influence a trait, there must be some degree of variation in the trait upon which natural selection can work.
5. **c** Since relatively few genes control much of development, changes in those genes can have dramatic, and often negative, effects on development. This limits the degree to which organisms can change through adaptive evolution.
6. **b** Because of these conflicting pressures, most adaptations are ecological compromises that match organisms with their environment well, but not perfectly.
7. **a** In the case of the male túngara frogs, the conflicting pressures are reproduction and predation.
8. **True** This example reveals one of the problems of a definition of the term *species* with a reproductive emphasis, because such a definition does not easily accommodate species such as these oaks, which can form fertile hybrids.
9. **d** Although poor reproductive performance by hybrids can be an isolating mechanism, the mere

production of hybrids actually links the species reproductively.

10. **e** All of these are potential problems with defining species based on reproduction.

11. **a** By definition, reproductive isolation means that there can be no flow of genes between the populations involved.

12. **b** Recall that ring species occur when populations loop around a geographic barrier, such as a mountain. Populations at either of the extreme ends of the barrier, while capable of interbreeding with members of intermediate populations, are incapable of interbreeding with each other.

13. **True** The East African cichlid fish discussed in this chapter are examples.

14. **f** allopatric speciation
 e domestication
 a polyploidy
 c speciation
 b hybrid
 d sympatric speciation

Conceptual Understanding

1. **a** Both are examples of the way natural selection shapes adaptive evolution, often in complex ways.

2. **c** Because female cichlids rely on coloration to choose their mates, they will begin to interbreed. The cloudy water would prohibit them from distinguishing their own species based on coloration, resulting in interbreeding and the loss of these species.

3. **b** The guppies that are found in the stream can be colorful because there are few predators. In a lake, there are more predators and colorful individuals will stand out. Therefore, the color of the transplanted fish will change from bright to drab relatively rapidly.

4. **e** Refer to Figure 18.1 in your textbook for this information.

5. **d** Reproductive isolation means that gene flow does not occur, which sets the stage for possible divergence of traits that eventually can lead to speciation.

6. **b** The ability of some species to reproduce fertile offspring through hybridization points out one of the limitations of a definition of *species* that is based on reproduction.

7. **d** Some of the other answers listed are frequently associated with speciation, but only reproductive isolation is essential.

8. **b** The exact distance that constitutes geographic isolation of populations depends on the type of locomotion used by a species. This distance must limit the gene flow between isolated populations.

9. **a** Reproductive isolation is necessary for speciation to occur. Likewise, natural selection must take place for adaptive evolution to occur.

10. **False** Polyploidy occurs with considerable frequency in plants, such that over half of all living plant species are descended from ancestors that underwent polyploid speciation. This mechanism is used infrequently in animals.

11. **c** Since all of these dog breeds can potentially reproduce with one another, they are best considered one species with many varieties.

12. **e** Although some natural speciation events can occur relatively rapidly (for example, over hundreds or a few thousand years), most appear to take hundreds of thousands or even millions of years to complete.

13. **e** All of the above have contributed to the sympatric speciation of these flies. Although they are not geographically isolated, they have become reproductively isolated over time.

14. **False** The distance needed for populations to become geographically isolated depends on the mode of locomotion of the organisms. Animals such as birds can travel much greater distances over a day than can snails, and therefore are more capable of reaching distant populations, facilitating gene flow.

15. **b** Finding long-lasting and effective treatments against disease organisms is made more difficult because these organisms are constantly adapting to defeat our treatments. Likewise, the process of domestication involves intense selective breeding for favorable traits that meet the needs of humans, thus leading to new adaptations for plants and animals.

CHAPTER **19**

The Evolutionary History of Life

Getting Started

Below are a few questions to consider prior to reading Chapter 19. These questions will help guide your exploration and assist you in identifying some of the key concepts presented in this chapter.

1. How many species of flowering plants are found on the continent of Antarctica?
2. What is macroevolution?
3. How long did it take for the ancestors of the whale to make the transition from life on land to life in water?
4. What role did the increase in oxygen concentration in the atmosphere play in the evolutionary history of life?
5. During what geologic time period did most of the major living animal phyla appear in the fossil record?
6. During what geologic time period did the supercontinent Pangaea begin to break up?
7. Approximately how many species are presently listed as endangered or threatened with extinction?
8. According to the fossil record, from which group of organisms did the mammals evolve?

A Guide to the Reading

The following concepts typically give students the most difficulty when exploring the content in Chapter 19 for the first time. For each concept, one or more references have been identified that may help you gain a better understanding of these potentially problematic areas.

Carbon Dating

Radioisotopes, radioactive forms of chemical elements, occur naturally in the environment. These unstable substances undergo the process of radioactive decay, transforming into a more stable version of the element. The decay of the radioisotope is accompanied by the release of radioactive energy that we can measure as radioactivity. The key to understanding how this process can be used to determine the age of fossils is realizing that not only do these compounds exist naturally, but the process of decay occurs naturally as well. One radioactive substance that is used to date fossils is carbon-14. Recall from Chapter 4 that the element carbon typically has an atomic weight of 12. Carbon-14, in contrast, has an atomic weight of 14 due to the presence of additional neutrons, which make the nucleus unstable. When carbon-14 undergoes decay, the additional neutrons are released, yielding a stable carbon atom with an atomic weight of 12. The release of neutrons during the decay process is measured as radioactivity. The time it takes for 1/2 the amount of carbon-14 (e.g. 5 out of every 10 atoms) in a sample to decay to carbon-12 is 5,730 years. Therefore, by knowing that all fossils originated with roughly the same percentage of carbon-14 present, and by measuring the amount of carbon-14 that remains in a fossil, scientists are able to estimate the age of the fossil with relative accuracy.

For more information on this concept, be sure to focus on:

- Section 19.1, The Fossil Record: A Guide to the Past
- Figure 19.1, Fossils through the Ages

Continental Drift

While it may seem impossible, Earth's continents are in constant motion. The movement of the continents that occurs over time is called *continental drift* and results from the sliding of "plates" that float on the surface of the earth's liquid core. This is how large land masses, such as the Pangaea super-continent that existed over 200 million years ago, have broken apart resulting in the continents as we know them today. Recall from Chapter 18 that the geographic isolation of species that occurs when land masses separate in this fashion can contribute to speciation as populations become reproductively isolated and have reduced gene flow.

For more information on this concept, be sure to focus on:

- Section 19.3, The Effects of Continental Drift
- Figure 19.7, Movement of the Continents over Time

Adaptive Radiation

Adaptive radiation refers to the process that occurs when a group of organisms "expands" to inhabit new habitats and ecological roles. Adaptive radiation often occurs after various groups of organisms go extinct, removing ecological competition for resources. With reduced competition or predation, surviving organisms are more free to expand to take advantage of the resources available in an ecosystem. In other cases, adaptive radiation may occur after an organism has acquired an adaptation that allows it to survive or reproduce more efficiently within an ecosystem. Adaptive radiations can occur on a small or a large scale, depending upon the supporting ecosystem.

For more information on this concept, be sure to focus on:

- Section 19.5, Adaptive Radiations: Increases in the Diversity of Life

Tying It All Together

Several concepts presented in this chapter build upon concepts presented in previous chapters and may also be revisited and discussed in greater detail in subsequent chapters, including:

Total Number of Species on Earth

- In Interlude A – How Many Species Are There on Earth?

Oxygen and Metabolism

- Chapter 7 – Section 7.2, Using Energy from the Controlled Combustion of Food

Natural Selection

- Chapter 16 – Section 16.2, Mechanisms of Evolution

Fossil Record

- Chapter 16 – in Section 16.4, Evolution is strongly supported by the fossil record

Adaptive Radiation of Galapagos Finches

- Chapter 16 – in Section 16.4, New species can be produced experimentally and their formation can be observed in nature

Speciation and Geographic Isolation

- Chapter 18 – in Section 18.4, Speciation often results from geographic isolation

Macroevolution

- In Interlude D – The Effects of Humans on Evolution

Practice Questions

Factual Knowledge

1. Which of the following continents was once a warm place covered with exotic plants and animals, but is now a cold desert where few organisms can survive?
   a. Antarctica
   b. North America
   c. Africa
   d. Pangaea
   e. Australia

2. Although only 1.5 million species have been described so far, estimates for the total number of species on Earth range from
   a. 2 to 3 million.
   b. 3 to 10 million.
   c. 3 to 30 million.
   d. 30 to 100 million.
   e. 100 million to 1 billion.

3. The first fossil evidence for life on Earth dates back to ________ years ago.
   a. 1.5 million
   b. 50 million
   c. 530 million
   d. 3.5 billion
   e. 4.6 billion

4. Which of the following is *not* a reason that the fossil record is incomplete?
   a. Decomposition often destroys organisms before they can fossilize.
   b. Evolution has not produced many organisms that fossilize well.
   c. The special circumstances needed for fossils to form are often not present.
   d. Geologic processes frequently destroy rocks before fossils are found.
   e. Locating fossils in rock formations is a difficult task.

5. Only rarely does the order in which organisms appear in the fossil record agree with our understanding of the evolutionary progression of life. (True or False)

6. Refer to Figure 19.1 in your textbook. Fossils such as those shown have been found in which of the following natural materials?
   a. sedimentary rock
   b. tree resin
   c. ice
   d. all of the above
   e. none of the above

7. Which of the following dates indicates the time when multicellular life first appeared on Earth?
   a. 4.6 bya
   b. 3.5 bya
   c. 650 mya
   d. 530 mya
   e. 65 mya

8. During the Cambrian explosion
   a. a large and relatively sudden increase in the diversity of life took place.
   b. continental drift resulted in one of the five mass extinctions.
   c. there was no occurrence, according to the evidence, of adaptive radiation.
   d. humans first began driving organisms to extinction.
   e. none of the above

9. Of the following organisms, which one is believed to have been the first eukaryote to colonize the land?
   a. mushrooms
   b. green algae
   c. bacteria
   d. fish
   e. amphibians

10. Continental drift has played a major role in macroevolution, in part because the movement of continents helps create the geographic isolation that promotes speciation. (True or False)

11. Approximately when in Earth's history did the fishes suddenly increase in diversity?
    a. 65 mya
    b. 205 mya
    c. 250 mya
    d. 365 mya
    e. 445 mya

12. One way that mass extinctions affect the history of life is by
    a. eliminating entire groups of organisms.
    b. reducing ecological and evolutionary opportunities for other organisms.
    c. causing continents to move and climates to change.
    d. inhibiting adaptive radiation.
    e. all of the above

13. Which of the following statements about the evolutionary history of life on Earth is *false*?
    a. Mass extinctions have played a key role in macroevolution.
    b. Continental drift has shaped the course of evolution for many organisms.
    c. Evolution above the species level differs from evolution of populations.
    d. The overall diversity of life has decreased through time.
    e. none of the above

14. All species that have highly beneficial adaptations will survive a mass extinction. (True or False)

15. Match each term with the best description.
    __ Cambrian explosion
    __ adaptive radiation
    __ fossils
    __ macroevolution
    __ mass extinction
    a. evolution above the species level
    b. process where groups of organisms expand to inhabit niches vacated by extinct groups
    c. period during which species numbers decline
    d. period during which species numbers increased dramatically
    e. preserved remains or imprints of organisms

Conceptual Understanding

1. The long-term, macroevolutionary changes in life that have occurred in Antarctica are best explained by
   a. continental drift.
   b. the Cambrian explosion.
   c. increases in atmospheric oxygen.
   d. fossilization.
   e. all of the above

2. In macroevolution, the focus is on
   a. changes that occur within individual populations.
   b. the overall pattern of evolutionary change through time.
   c. the geological processes that cause continental drift.
   d. ecological circumstances that reduce competition.
   e. how fossils are formed.

3. A key factor that made eukaryotic life possible was
   a. continental drift.
   b. the Cambrian explosion.
   c. the colonization of land.
   d. reaching a threshold level of atmospheric oxygen.
   e. the elimination of bacterial competitors.

4. Which of the following macroevolutionary events occurred before the emergence of photosynthetic bacteria?
   a. Cambrian explosion
   b. first multicellular life
   c. first eukaryotes
   d. extinction of the dinosaurs
   e. none of the above

5. The emergence of humans, the extinction of the dinosaurs, and the colonization of land by plants and animals all
   a. were part of mass extinction events.
   b. occurred before the Cambrian explosion.
   c. are significant macroevolutionary events in the history of eukaryotes.
   d. all took place prior to the emergence of multicellular life.
   e. none of the above

6. Fossilized human remains are often found in the same rocks that contain dinosaur fossils. (True or False)

7. Which of the following is *not* one of the challenges that animals had to face during the colonization of land?
   a. supporting the body
   b. reproduction
   c. conserving ions and water
   d. becoming multicellular
   e. movement

8. Of the following sequences, which one best represents the evolution of vertebrates?
   a. reptiles - amphibians - fish - birds - mammals
   b. amphibians - fish - reptiles - birds - mammals
   c. fish - reptiles - amphibians - mammals - birds
   d. fish - amphibians – reptiles - mammals - birds
   e. birds - mammals - reptiles - amphibians - fish

9. One of the major side effects of continental drift is ________, which in turn leads to ________.
   a. fossilization; adaptive radiation
   b. microevolution; climate change
   c. climate change; the extinction of many species
   d. mass extinction; a reduction in the amount of atmospheric oxygen
   e. release from competition; mass extinction

10. Of the five major mass extinctions, the one that had the most dramatic effect on macroevolution occurred
    a. 65 mya.
    b. 205 mya.
    c. 250 mya.
    d. 365 mya.
    e. 440 mya.

11. Adaptive radiations can be caused when species are released from competition. A dramatic example of this occurred 65 mya when ________ emerged from the decline of ________.
    a. reptiles; amphibians
    b. mammals; dinosaurs
    c. insects; plants
    d. birds; mammals
    e. none of the above

12. The appearance of groups of animals in the fossil record does not match the order based on DNA and morphological evidence. (True or False)

13. Which of the following statements regarding the differences between mammalian and reptilian jaws and teeth is *false*?
    a. Extra muscles present in the cheek region allow for increased strength and precision of the mammalian jaw.
    b. The temporal fenestra present in modern mammals allows for the attachment of an extra muscle to the jaw.
    c. The Therapsids were the first group to develop a temporal fenestra.
    d. Modern mammals have highly specialized teeth, which differ depending upon location within the jaw.
    e. The enlarged jaw typical of mammalian species first appeared in the cynodonts.

14. Refer to Figure 19.8 in your textbook. The large increase in diversity that began 65 mya is mainly the result of
    a. adaptive radiation.
    b. continental drift.
    c. new fossil formation.
    d. high gene flow between populations.
    e. increased oxygen in the atmosphere.

15. A mass extinction has just occurred, wiping out 85 percent of all land animals. What are the key effects on the diversity of life after this mass extinction?
    a. One type of organism from all major groups will survive.
    b. New ecological and evolutionary opportunities will become available to those organisms that survive.
    c. Biodiversity will not change.
    d. all of the above
    e. none of the above

Related Activities

- Visit a few of the Web sites hosted by various natural history museums in the United States (for example, Smithsonian, Carnegie, American Museum of Natural History, Field Museum of Chicago). Research the concept of macroevolution on these sites and compose a one-page summary describing how the exhibits at these locations illustrate the concept of macroevolution.
- Antarctica today is a biologically stark continent when compared with the period of time when it was part of Pangaea and harbored abundant life. Consult a geology or biology reference source about the evolutionary history of Antarctica. Then create an evolutionary timeline that shows how Antarctica drifted from nearer the equator to its current position over the South Pole. Be sure to include on your time line major evolutionary events of relevance to the organisms that lived on and around Antarctica.
- Review the article "When Bats and Humans Were One and the Same" in the Biology in the News section at the end of the chapter. This article describes how DNA preserved in fossils can be used to determine evolutionary relationships to modern organisms and how DNA sequences from extinct organisms may be "predictable" based upon these evolutionary relationships. Search the World Wide Web for an example of how scientists are using DNA technology to study evolutionary history. Compose a one-page summary describing a few of the studies you have identified.

Answers and Explanations

Factual Knowledge

1. **a** The macroevolutionary change that has occurred in Antarctica is stunning and is an excellent example of the effect of continental drift.
2. **c** Some estimates go as high as 100 million, but the consensus view is 3 to 30 million species.
3. **d** The age of the Earth is 4.6 billion years, but the first fossil life does not appear until 3.5–4 bya.
4. **b** Many of the organisms that have lived in the past could have potentially become fossils; however, because of the reasons given in answers a, c, d, and e, they did not produce fossils or the fossils were never found.
5. **False** The two are often closely matched, providing strong support for the predictions made by evolutionary theory.
6. **d** Fossils can be found in a variety of materials, including those listed.
7. **c** There was a long period of time between the first appearance of life (3.5 bya) and the evolution of multicellular organisms (650 mya).
8. **a** Although the reasons for the Cambrian explosion remain unclear, it was one of the most dramatic examples of increased diversity in the history of life.
9. **b** Bacteria were actually the first organisms to live on land, but they are prokaryotes.
10. **True** When continents separate, populations become isolated geographically. This reduces gene flow and promotes speciation.
11. **e** In many ways, the increase in fish diversity 445 mya began the serious evolution of the vertebrate classes.
12. **a** The second and fourth answers tend to be increased by mass extinctions, while the third answer causes mass extinctions.
13. **d** Overall diversity has increased, and dramatically so during the past 200 million years. See Figure 19.8 in your textbook.
14. **False** Survival of a mass extinction may be a purely random event and having highly beneficial adaptations may not ensure survival.
15. **d** Cambrian explosion
    b adaptive radiation
    e fossils
    a macroevolution
    c mass extinction

Conceptual Understanding

1. **a** On Antarctica, continental drift isolated all but the flying and swimming organisms. When drastic climate changes occurred, the majority of those organisms could not adapt and went extinct.
2. **b** Macroevolution is concerned with all of these ideas in some way, but the main focus is on the overall pattern of change through time.
3. **d** Eukaryotes are larger than prokaryotes, and so need more oxygen to survive. It wasn't until the proper atmospheric oxygen threshold was reached that eukaryotic life was possible.
4. **e** The evolution of photosynthetic bacteria occurred very early in the history of life, pre-dating all of the events listed. See the timeline in Figure 19.3 of your textbook.
5. **c** Recall that macroevolution focuses on the overall pattern of change over time. These events are all significant macroevolutionary events that have occurred in the history of eukaryotes on earth.

6. **False** Humans and dinosaurs are separated on the geologic timescale by over 50 million years of evolution.
7. **d** The first land animals would have already been multicellular organisms.
8. **d** Vertebrate life first began in water with fish, then partially moved to land with amphibians, and subsequently became specialized for the terrestrial world with reptiles, mammals, and birds. Refer to the geologic timescale in Figure 19.3.
9. **c** As continents move, climates change because of distance from the equator and shifts in ocean currents. These changes can force species into extinction.
10. **c** The great Permian mass extinction saw enormous numbers of species and other taxonomic groups completely disappear, especially from the marine environment.
11. **b** The mass extinction of 65 mya eliminated the dinosaurs and removed a major source of competition for mammals, whose numbers and varieties then increased.
12. **False** The fossil record agrees very well with the genetic and morphological evidence to provide support for evolution.
13. **c** Recall that the temporal fenestra was an early adaptation that first appeared in ancestral reptiles (*Haptodus*). Review Figure 19.12 for additional information.
14. **a** The increase is an adaptive radiation caused by the removal of many species, including the dinosaurs, through a mass extinction event 65 mya.
15. **b** When mass extinctions occur, entire groups of organisms are lost. This tends to open up more ecological and evolutionary opportunities to those organisms that do survive.

INTERLUDE D Humans and Evolution

Getting Started

Below are a few questions to consider prior to reading Interlude D. These questions will help guide your exploration and assist you in identifying some of the key concepts presented in this interlude.

1. From what German valley did the Neandertals originate?
2. What is the scientific name for the earliest known hominid?
3. What are some of the tools that were developed during the early evolution of the hominids?
4. What is the multiregional hypothesis?
5. What has happened to the size of the human brain over the past 75,000 years?
6. Approximately how much does the development of drug resistant diseases cost society each year?
7. What do scientists expect to happen to the extinction rate over the next 100 years?
8. What is eugenics?

A Guide to the Reading

The following concepts typically give students the most difficulty when exploring the content in Interlude D for the first time. For each concept, one or more references have been identified that may help you gain a better understanding of these potentially problematic areas.

Multiregional vs. Out-of-Africa Hypotheses

The many fossil remains found in Africa, China, and Europe that appear to be intermediate between *Homo erectus* and *Homo sapiens* have prompted scientists to develop different hypotheses regarding the origin of modern humans. The out-of-Africa hypothesis suggests that anatomically modern humans evolved from a single ancestral line in Africa (archaic *Homo sapiens*) and later spread to other continents, replacing the native hominid populations. The key to this hypothesis is realizing that there would have been little or no gene flow between the archaic *Homo sapiens* and native species, maintaining genetic isolation. In contrast, the multiregional hypothesis provides a more mixed view of the origin of modern humans. In this hypothesis, it is believed that archaic *Homo sapiens* evolved in parallel from *Homo erectus* populations that were present in Europe, Africa, and Asia. As a result, regional differences between human populations in different areas of the world would have developed early in the evolutionary process. It is also believed that the adaptations that characterize modern humans would have been shared between these populations as a result of gene flow. Currently, evidence exists to support each hypothesis. As a result, scientists are not certain which hypothesis is, in fact, correct.

For more information on this concept, be sure to focus on:

- In Interlude D, The Origin and Spread of Modern Humans
- Figure D.9, The Origin of Anatomically Modern Humans

The Sixth Mass Extinction

As discussed in the interlude, we may currently be in the midst of a sixth mass extinction, one that is caused mainly by humans. The environmental impact of human activities is predicted to have the most significant effect in terms of extinction that has ever occurred in the history of the Earth. Global warming, pollution, and depletion of natural resources and habitat are causing species to go extinct at rates that may soon dwarf the rates seen in previous mass extinction events. In addition, human activities are also influencing the microevolution of various species that may have a detrimental impact on the health of numerous species as well as our own. Examples include the evolution of pesticide and antibiotic resistance due to the widespread and indiscriminate application of pesticides and antibiotics. In addition to the detrimental effects on the health of humans, these developments have also caused (and will continue to cause) widespread economic trouble.

For more information on this concept, be sure to focus on:

- In Interlude D, People have profound effects on microevolution in other species
- In Interlude D, People also alter macroevolutionary patterns
- Figure D.11, Directional Selection for Pesticide Resistance
- Figure D.12, Human Effect on the Landscape

Tying It All Together

Several concepts presented in this chapter build upon concepts presented in previous chapters and may also be revisited and discussed in greater detail in subsequent chapters, including:

Alleles

- Chapter 10 – Section 10.2, Gene Mutations: The Source of New Alleles

Mass Extinctions

- In Interlude A – The Beginnings of a Present-Day Mass Extinction
- Chapter 19 – Section 19.4, Mass Extinctions: Worldwide Losses of Species

Natural Selection

- Chapter 16 - Section 16.2, Mechanisms of Evolution
- Chapter 17 - Section 17.7, Natural Selection: The Effects of Advantageous Alleles

Gene Flow

- Chapter 17 - Section 17.5, Gene Flow: Exchanging Alleles Between Populations

Genetic Drift

- Chapter 17 - Section 17.6, Genetic Drift: The Effects of Chance

Microevolution

- Chapter 19 - in Section 19.7, Macroevolution differs from the evolution of populations

Practice Questions

Factual Knowledge

1. Which of the following statements about hominids is *false*?
   a. They are mammals.
   b. They are primates.
   c. They have relatively large brains.
   d. They are quadrupedal.
   e. They include the Neandertals.
2. Which of the following is *not* a characteristic that humans share with the other great apes?
   a. use of tools
   b. communication with symbolic language
   c. acts of deception
   d. sense of self-awareness
   e. none of the above
3. Which of the following would *not* be considered one of the great apes?
   a. human
   b. gorilla
   c. orangutan
   d. chimpanzee
   e. tree shrew
4. Which of the following would *not* be considered an advantage for an organism that evolved the ability to walk upright?
   a. freeing hands to carry objects
   b. freeing hands to use tools
   c. improving line of sight
   d. greater mobility in trees
   e. none of the above
5. According to the fossil record, which of the following dates is not matched correctly with an important event in primate evolution?
   a. 65 mya—first primates appear
   b. 50 mya—first hominids appear
   c. 3 to 4 mya—first *Australopithecines* appear
   d. 2 to 3 mya—first *Homo* appears
   e. All of the above are correctly matched.
6. When hominids made the transition from life in the trees to life on the ground, which of the following also happened?
   a. Brain size decreased.
   b. Methods of locomotion began to change.
   c. The fully opposable big toe first appeared.
   d. Daytime vision was reduced.
   e. none of the above
7. Fossils of the earliest hominids (5–7 mya) clearly indicate that they
   a. walked upright.
   b. had relatively small brains.
   c. had the ability to make tools.
   d. had skulls and teeth that were apelike.
   e. all of the above
8. The evolutionary origin of the primates dates back 65 million years to a group of nocturnal, tree-dwelling, insectivorous small mammals similar to modern tree shrews. (True or False)
9. In which of the following mammals would you expect to find a pronounced arch in the foot?
   a. modern human
   b. gorilla
   c. monkey

d. lemur
e. none of the above

10. Of the following hominid species, which one made the most sophisticated tools?
a. *Homo erectus*
b. *Homo habilis*
c. *Homo sapiens*
d. *Australopithecus afarensis*
e. *Ardipithecus ramidus*

11. Fossil evidence for the existence of the genus *Homo* dates back to
a. 65 mya.
b. 35 mya.
c. 15 mya.
d. 4.2 mya.
e. 2.4 mya.

12. Overall, the available evidence suggests that the best explanation for the origin of modern humans is the ________ hypothesis.
a. multiregional
b. out-of-Africa
c. upright-posture
d. trend-toward-big-brains
e. chimpanzees-are-our-closest-relatives

13. The brain size of the hominids has continued to grow over the past 75,000 years. (True or False)

14. Which of the following statements regarding human effects on the microevolution of other species is *false*?
a. Insects rapidly develop resistance to insecticides.
b. Disease-causing viruses rapidly develop resistance to drugs used to combat them.
c. Due to intense fishing, various species have evolved thinner bodies.
d. Salmon hatcheries are producing fish that are much larger than those found in nature.
e. none of the above

15. Match each term with the best description.
__ eugenics
__ multiregional hypothesis
__ *Homo floresiensis*
__ hominids
__ primates
a. one order within the class Mammalia
b. reproduction of humans with "favorable" traits
c. simultaneous evolution of hominid groups on different continents
d. miniature species of *Homo*
e. today includes only one living species

Conceptual Understanding

1. Refer to Figure D.2 in your textbook. Of the modern species shown, which one is most like the ancestral early primates?
a. gorilla
b. human
c. chimpanzee
d. orangutan
e. none of the above

2. If you could go back in time and compare the very first primates with modern lemurs, you would find that both would
a. have large brains similar in size to modern chimpanzees.
b. walk upright when moving on the ground.
c. be well-adapted for life in the trees.
d. have large numbers of offspring.
e. use tools.

3. Recent evidence suggests that modern humans are the direct descendants of Neandertals. (True or False)

4. Which of the following anatomical characteristics did some of the earliest hominids have?
a. enlarged brain size
b. quadrupedal locomotion
c. humanlike teeth
d. partially opposable big toe
e. all of the above

5. In human evolution, which of the following would have occurred *before* the switch to walking upright?
a. tool use
b. mastery of fire
c. spoken language
d. enhanced grasping ability
e. loss of opposable big toe

6. Imagine that you visit a museum collection of hominid fossils and carefully examine the skeletons. Which of the following species would show anatomical evidence of upright walking?
a. *Homo erectus*
b. *Homo habilis*
c. *Australopithecus afarensis*
d. *Australopithecus africanus*
e. all of the above

7. You are shown the skeletal remains of the *Homo erectus* species. What features tell you that it is *not* from a modern human?
a. relatively large brain size
b. flattened face

c. type of teeth
d. taller and more robust
e. all of the above

8. Which of the following hominids would *not* have overlapped and co-existed with *Homo sapiens* at one time or another in the history of earth?
   a. *Homo erectus*
   b. *Homo floresiensis*
   c. Neandertals
   d. *Australopithecus afarensis*
   e. none of the above

9. Suppose that a new fossil hominid is discovered. What criterion can be used to determine where in the human evolutionary lineage this new fossil belongs?
   a. brain size
   b. tooth structure
   c. foot structure
   d. evidence of tool use
   e. all of the above

10. Over the greater span of hominid evolution, the ratio of brain size to body weight has generally ________ During the past 75,000 years, however, the brain size of Homo appears to have ________
   a. decreased; increased
   b. increased; decreased
   c. decreased; stayed the same
   d. increased; stayed the same
   e. stayed the same; increased

11. Compared to the tool use of Neandertals, the sophisticated technology and complex social systems of modern humans result from the fact that modern people have much larger brains than did Neandertals. (True or False)

12. Genetic drift is thought to have played a large role in shaping the evolution of early human populations. In human populations today, genetic drift is ________ important, because ________.
   a. very; many human populations are still small
   b. very; many human populations are still isolated
   c. very; of its effect on genes related to brain size
   d. not important; gene flow between populations is low
   e. not important; the human population is now large and mobile

13. One of the main lines of evidence against the multiregional hypothesis of human evolution is that there appeared to be high levels of gene flow between the different Homo groups. (True or False)

14. Which of the following statements regarding the role of natural selection in current human evolution is *false*?
   a. Natural selection plays a larger role in human evolution due to our reliance on tools and technology.
   b. Natural selection plays a role in the evolution of disease resistance in human populations.
   c. Natural selection now plays a lesser role than gene flow in shaping human evolution.
   d. Natural selection now plays a greater role than genetic drift in shaping evolution.
   e. none of the above

15. Which of the following human effects on the environment is most likely to contribute to the greatest mass extinction event in the history of the earth?
   a. pollution
   b. destruction of the earth's tropical forests
   c. indiscriminate use of pesticides
   d. destruction of continuous habitats
   e. introduction of genetically engineered species to new ecosystems

Related Activities

- Imagine that you have been asked to talk to a group of middle school children about the scientific explanation of human origins. Compose a one-page summary explaining how you would describe this event to this group. Be sure to state which examples you would use to best illustrate the story.
- Reflect on the scientific explanation of human origins, and then compare those ideas to your own religious and philosophical views about humans and their role in the natural world. Compose a one-page essay on how well the two match, and the degree to which they do not. In your essay, discuss how you might reconcile the differences.
- Review the Biology Matters article "Humans: The World's Dominant Evolutionary Force," which appears at the end of this interlude. In a one-page essay, describe what you think are the most beneficial and harmful macroevolutionary effects that humankind has directly influenced. For your harmful examples, indicate strategies that you feel would help rectify these situations.

Answers and Explanations

Factual Knowledge

1. **d** Recall that the hominid designation is reserved for humans and their extinct humanlike ancestors. All hominids characteristically have an upright, bipedal walking posture.

2. **e** All of the listed characteristics are shared between humans and the other great apes.
3. **e** Tree shrews are small insectivorous mammals that resemble what is thought to be the mammalian ancestor from which the primate order first evolved.
4. **d** Recall that the ability to walk upright was a hominid adaptation that enabled them to better adapt to life on land, as opposed to trees.
5. **b** The first hominids appeared about 6 mya.
6. **b** Life on the ground set the stage for upright walking, which quickly became the dominant mode of locomotion in hominids.
7. **e** All of these traits are characteristic of the first hominids.
8. **True** Although the fossil record for early primates is sketchy, an animal that resembles the modern tree shrew is believed to be the mammal from which primates emerged 65–80 mya.
9. **a** The upright walking of humans and other hominids made the foot arch an adaptive advantage. Since the other primates don't walk upright, their feet do not show this adaptation.
10. **c** Both modern and archaic versions of *Homo sapiens* clearly made (and still make) the most sophisticated tools in the animal kingdom.
11. **e** The oldest specimens of *Homo habilis* date to this time.
12. **b** The first answer is the only other serious choice, but the multiregional hypothesis has recently been called into question because it does not appear that the extensive gene flow required to make the hypothesis feasible would have been possible with widely spread populations of early *Homo*.
13. **False** The average brain size for modern humans (1,350 cubic centimeters) is actually smaller than that observed in fossils of anatomically modern humans (1,500 cubic centimeters).
14. **d** Natural selection is producing males that are smaller than those found in the wild since smaller males survive better in hatcheries.
15. **b** eugenics
    c multiregional hypothesis
    d *Homo floresiensis*
    e hominids
    a primate

Conceptual Understanding

1. **d** Based on the evolutionary tree shown, the orangutan diverged roughly 12–16 mya and would be most closely related to the ancestral early primates.
2. **c** All the other features offered as answers either are not characteristics of primates in general or appeared well after the emergence of the first primates.
3. **False** The Neandertals were an advanced type of archaic *Homo sapiens* that co-existed with humans in western Asia for some 80,000 years.
4. **d** See Figure D.4 in your textbook for a diagram of the foot bones of the early hominids. The partially opposable big toe allowed these groups to continue to use trees some of the time.
5. **d** Enhanced grasping ability was a very early primate adaptation for life in the trees; therefore, it would have predated the other hominid characteristics.
6. **e** All of these are hominids, and all hominids walked upright to some degree.
7. **d** The features of *Homo erectus* are very similar to those of modern humans. They were, in general, taller and more robust than modern humans, however.
8. **d** Fossil evidence suggests that *Homo sapiens*, *Homo erectus*, *Homo floresiensis*, and the Neandertals all inhabited various regions of the Earth as recently as 30,000 years ago.
9. **e** All of these hominid characteristics have been influenced by evolution, and so they serve as legitimate bases for comparison of the new fossil to existing fossils.
10. **b** Dramatic increases in the ratio of brain size to body size are seen when all of human evolution is considered, but when comparisons are made between absolute brain size of Neandertals and modern humans, brain size actually decreases.
11. **False** Our brain is actually somewhat smaller than that of Neandertals, and so we must look elsewhere to explain the behavioral differences between these groups of hominids.
12. **e** Because there are so many humans today, traveling to all corners of the globe and interbreeding with each other, gene flow is high and the impact of genetic drift is thus low.

13. **False** It appears that there was little gene flow between different populations of early humans.

14. **a** Due to humankind's ability to craft tools and technology to overcome limitations (such as poor vision), natural selection now plays a less significant role in shaping human evolution.

15. **b** While all of the events listed are contributing to the extinction of thousands of species, perhaps the most devastating is the destruction of the tropical rainforests of the world. Since over 50 percent of the Earth's species live in these forests, their destruction would lead to the catastrophic extinction of over half of the Earth's existing species.

CHAPTER 20 The Biosphere

Getting Started

Below are a few questions to consider prior to reading Chapter 20. These questions will help guide your exploration and assist you in identifying some of the key concepts presented in this chapter.

1. What is the relationship between the zebra mussel and cargo ship ballast water?
2. What is the estimated annual impact on the U.S. economy attributable to introduced species?
3. Why are the Earth's equatorial regions so much warmer than its polar regions?
4. Why are most deserts located near 30° latitude?
5. What best explains why water temperatures on the Pacific coast of the United States are so much cooler than on the Atlantic coast?
6. What two climatic factors most strongly influence the development of a terrestrial biome?
7. Terrestrial biomes are named after the dominant vegetation of the region; how are aquatic biomes named?
8. What two factors most strongly influence the development of an aquatic biome?

A Guide to the Reading

The following concepts typically give students the most difficulty when exploring the content in Chapter 20 for the first time. For each concept, one or more references have been identified that may help you gain a better understanding of these potentially problematic areas.

The Importance of Ecology

Chapter 20 begins with the question "Why Is Ecology Important?" Although a complete answer is beyond our current understanding, it has become increasingly evident that people are dependent on the continuous operation of the biosphere's natural systems. Unfortunately, human activity is changing the biosphere in many ways, some of which may not be at all positive. Introduced species clearly illustrate this concern. Whether introduced intentionally or accidentally, plant and animal species from other regions may expand uncontrollably, displace native species, and reduce the usefulness or dependability of natural systems. Current estimates for economic damage related to introduced species are near $120 billion annually for the United States alone. Once it is established, removing an introduced species becomes all but impossible. Yet banning all future introductions may not be in our best interests either. More than 98 percent of the food currently produced in the United States comes from introduced crops. Plant species with potentially valuable uses are routinely discovered, and, despite our best efforts, accidental introductions will continue. A more complete understanding of ecology can help us understand the consequences of these and other environmentally damaging events (such as the dispersal of toxic chemicals), direct an appropriate response to current problems, and perhaps prevent future ones from occurring.

For more information on this concept, be sure to focus on:

- Section 20.1, Why Is Ecology Important?

Interactions with the Environment

Throughout this unit, you will be repeatedly exposed to the idea that the interactions between living organisms and the environment are reciprocal. Biologists have known for hundreds of years that the physical environment influences the appearance and distribution of plants and animals. Even without a formal education, almost everyone you know can explain why polar bears don't live at the equator. What's become increasingly clear is the influence living organisms have on the physical environment. This is illustrated by the example of nitrogen availability given in your text. Nitrogen is used by plants in only two molecular forms. When these forms are reduced in the soil, plant growth slows or ceases entirely. Since different plants often have differing nitrogen needs, any particular plant may become rare or more common when nitrogen declines. Thus the levels of soil nitrogen strongly influence the pattern of plant growth within any particular area. Some bacteria convert atmospheric nitrogen into the forms that support plant growth. The presence of these bacteria will elevate nitrogen levels in the soil, providing some plants with a competitive advantage over others. Two physical localities, identical in all respects, but differing in soil bacteria, may look entirely different.

For more information on this concept, be sure to focus on:

- Section 20.2, Interactions with the Environment

Climate Has a Large Effect on the Biosphere

The Earth's surface is not uniformly heated. Sunlight strikes the equator and the Earth's tropical regions directly, but strikes the poles and higher latitudes at a slanted angle. Although these angles change somewhat with the seasons, the tropics experience the least variation, providing a relatively warm, stable climate throughout the year for the organisms living there. Polar regions cycle through relatively short, mild summers to the bitter cold of winter. Such extremes over the course of a year make life in the polar regions much more challenging than life in the tropics. Differences in the temperature of the Earth's surface heat or cool the adjacent atmosphere. In tropical regions the moist air warms and rises, but expands and cools when doing so. Because cool air has a lower capacity to hold moisture, rain will eventually fall. In addition, the loss of moisture has also reduced the air's weight and so it is unable to sink through the moist air rising from the surface beneath it. The air is first displaced to the north and south, eventually sinking at about 30° latitude. As it descends the air is compressed, causing it to warm and gain moisture from the Earth's surface as it moves back toward the equator. By the time it reaches the equator the air is warm and moist and the cycle is repeated. This cyclical circulation of air is termed a *convection cell.* Two convection cells exist in each hemisphere. A polar cell, with air descending at the pole and flowing to about 60° latitude, where it rises, and the tropical cell just described. The horizontal surface component of each cell produces the winds that most directly influence the biosphere. These winds are relatively consistent in the tropics and polar regions, but somewhat variable in the temperate regions between 30° and 60° latitude. The winds, however, do not move directly north or directly south as the orientation of the convection cells would suggest. Influenced by the rotation of the Earth, they curve to the right in the northern hemisphere and to the left in the southern. In the northern hemisphere the surface winds blowing south toward the equator turn to the right, appearing to come from the east; they are called easterlies. Surface winds traveling poleward are also turned to the right, appear to come from the west, and are called westerlies.

For more information on this concept, be sure to focus on:

- In Section 20.3, Incoming solar radiation shapes climate
- In Section 20.3, Global movements of air and water shape climate
- Figure 20.3, Earth Has Four Giant Convection Cells
- Figure 20.4, Global Patterns of Air Circulation

Biomes

The biosphere can be divided into several terrestrial and aquatic life zones termed *biomes.* Seven terrestrial and eight aquatic biomes are commonly recognized. Terrestrial biomes are named for the dominant forms of vegetation within each region. Examples include forests and grasslands. Rainfall and temperature are the primary factors that influence biome development. When rainfall is abundant a forest can form. Recall that rainfall is high at the equator and at 60° latitude. Forests exist in both locations, but the specific form of forest, tropical rainforest or boreal forest, for example, is determined primarily by average temperature. Because water is relatively thermally stable, aquatic biomes are less influenced by temperature than terrestrial biomes. Thus the annual temperature range for most aquatic biomes is small. Climate, however, still significantly influences most aquatic biomes. As was true of air, the temperature of water directly affects its density. Vertical mixing can occur in lakes and the ocean when surface waters cool and sink. In lakes, such mixing may move oxygen-rich surface water to the bottom and displace nutrient-rich water toward the surface. Marine aquatic biomes may be affected by temperature in additional ways. High temperatures may evaporate water from a marine biome, concentrating salt and increasing salinity, while rainfall may dilute saltwater and lower salinity. While the situations described above tend to be localized, physical conditions may sometimes alter a significant portion of an entire ocean basin. Consider the El Niño events that develop periodically off the western coast of South America. Unusually warm ocean water floats over the top of cooler nutrient-rich water, preventing it from reaching the surface. Without the replenishment of nutrients the abundance of marine life declines spectacularly.

For more information on this concept, be sure to focus on:

- In Section 20.4, The biosphere can be divided into biomes
- In Section 20.5, Aquatic biomes are influenced by terrestrial biomes and climate
- In Section 20.5, Aquatic biomes are also influenced by human activity
- Figure 20.12, El Niño Events

Tying It All Together

Several concepts presented in this chapter build upon concepts presented in previous chapters and may also be revisited and discussed in greater detail in subsequent chapters, including:

The Importance of Ecology

- Chapter 20 - Section 20.6, Human Population Growth: Surpassing the Limits?
- In Interlude E, The Current Human Impact Is Not Sustainable

Interactions with the Environment

- Chapter 23 - Section 23.3, Recovery from Disturbances

Climate Has a Large Effect on the Biosphere

- Chapter 23 - in Section 23.2, Communities change as climate changes

Biomes

- Chapter 24 - in Section 24.2, The rate of energy capture varies across the globe

Practice Questions

Factual Knowledge

1. One major difference between weather and climate is that
   a. weather changes rapidly but climate is predictable.
   b. climate shifts rapidly but weather is predictable.
   c. weather is cyclical.
   d. climate is the weather during a short interval.
   e. none of the above

2. Seasons are influenced by
   a. the tilt of Earth on its axis.
   b. the amount of solar radiation reaching Earth's surface.
   c. Earth's movement around the sun.
   d. all of the above
   e. none of the above

3. The term *convection cell* refers to
   a. the rising of cool, dry air and the descending of warm, moist air.
   b. a factor that contributes to wind patterns.
   c. the rising of warm, moist air and the descending of cool, dry air.
   d. both a and b
   e. both b and c

4. Deserts are most likely to occur
   a. at 30° above and below the equator.
   b. in the rain shadows of mountains.
   c. where cool, dry air descends.
   d. all of the above
   e. none of the above

5. Places like London, England, are generally warmer than places of the same latitude in North America because
   a. the English Channel acts as insulation.
   b. the Gulf Stream transfers heat from the tropics.
   c. sunlight strikes the eastern hemisphere more directly.
   d. all of the above
   e. none of the above

6. Biomes are generally uniform in their species distribution. (True or False)

7. The makeup of any given biome is affected by which of the following factors?
   a. climate
   b. the species residing there
   c. geography
   d. human impact
   e. all of the above

8. Direct exclusion of a species from a biome means that
   a. the species cannot survive the climate of the biome.
   b. the species can survive the climate but is outcompeted by species better suited to the climate.
   c. human intervention has prevented the species from establishing itself.
   d. all of the above
   e. none of the above

9. In regions of hot temperatures and wet climate, you will most likely find ________ biomes, whereas in regions of hot temperatures and dry climate, you will find ________ biomes.
   a. desert; tropical
   b. temperate; arid
   c. tropical; desert
   d. arid; temperate
   e. tundra; chaparral

10. Terrestrial biomes are classified according to which of the characteristics below?
    a. physical characteristics of the environment
    b. dominant vegetation living there
    c. human impact on the biome
    d. all of the above
    e. none of the above

11. Aquatic biomes of the same type are very similar in terms of the species distributed within them. (True or False)

12. Terrestrial and aquatic biomes have little effect on each other. (True or False)

13. El Niño affects the coast of Peru by
    a. changing the flow of the current.
    b. warming the waters off the coast.
    c. changing the distribution of species off the coast.
    d. influencing weather patterns.
    e. all of the above

14. Match each term with the best description.
    __ biosphere
    __ ecosystem
    __ habitat
    __ ecology
    __ biome
    a. the study of interactions between organisms and their environment
    b. different species in an area, plus the environment in which they live
    c. all of the organisms and the environments in which they live
    d. one of several major terrestrial or aquatic life zones
    e. characteristic place where each organism lives

15. The hole created in the ozone layer by chlorofluorocarbons is problematic because
    a. heat from the Earth is escaping through the hole.
    b. ozone protects us from the sun's ultraviolet light.
    c. it leads to global warming.
    d. all of the above
    e. none of the above

Conceptual Understanding

1. Why is the study of ecology becoming increasingly important?
   a. Humans are increasingly affecting the environment.
   b. We truly understand very little about how organisms and their environments interact.
   c. Human population growth is putting more and more biomes at risk.
   d. all of the above
   e. none of the above

2. Occasionally the Gulf Stream follows a more southerly path; how might this affect the climate of Northern Europe?
   a. greater temperature ranges with hotter summers and colder winters
   b reduced temperature ranges with warmer winters and cooler summers
   c. a general cooling resulting in cooler winters and summers
   d. a general warming resulting in warmer winters and summers
   e. The Gulf Stream has no appreciable effects on the climate of Northern Europe.

3. You have been teleported to a terrestrial region of Earth you have never visited before. You determine which biome you are in by noting the
   a. climate.
   b. dominant vegetation living there.
   c. most obscure species you can find.
   d. both a and b
   e. both a and c

4. Climate events like El Niño are caused by changes in ocean currents. This indicates that ocean currents have a ________ impact on ________ climate.
   a. small; regional
   b. small; global
   c. large; global
   d. large; regional
   e. none of the above

5. You are in an area in which the water fluctuates between very salty and not so salty. You look around and see that the region is located at the place where a river meets the ocean. You are most likely in a(n) ________ biome.
   a. estuarine aquatic
   b. river aquatic
   c. tundra terrestrial
   d. intertidal aquatic
   e. none of the above

6. A friend writes to you and tells you that his community is very cold and there are many days of total darkness. They haven't had snow in quite some time, but the permafrost layer is very deep. Your friend is writing to you from the
   a. chaparral.
   b. boreal forest.
   c. tundra.
   d. desert.
   e. none of the above

7. Overfishing is one way humans affect aquatic biomes. The most serious aspect of overfishing is
   a. the oil and gas pollution associated with boat fuel tanks.
   b. the introduction of live bait into the biome.
   c. its potential to change the species distribution within the biome.
   d. noise pollution from boat motors.
   e. none of the above

8. The eastern side of the Rockies tends to receive much less rainfall than the western side because of
   a. convection currents.
   b. the rain shadow effect.
   c. human impact.
   d. all of the above
   e. none of the above

9. Equatorial regions are tropical because
   a. the sun strikes them more directly than it does other parts of the Earth.
   b. hot moist air rises, then cools and causes rain.
   c. large bodies of water provide ample moisture.
   d. all of the above
   e. none of the above

10. The biosphere depends on
   a. interactions among all organisms and the environment.
   b. humans living responsibly within it.
   c. physical factors such as climate.
   d. all of the above
   e. none of the above

Related Activities

- Many biological supply houses sell what they call "mini-biospheres," enclosed habitats that sustain a number of different species indefinitely. Using the World Wide Web or a supply house catalog, find out more about these mini-biospheres. Compare and contrast them to the biosphere.
- Investigate how human impact has changed distribution of biomes over the past 100 years.
- Select a biome of interest. Discuss how variable that biome is with respect to species distribution. Discuss which factors affect its species distribution.
- Investigate the greenhouse effect. What are its causes and what do scientists predict its consequences will be?

Answers and Explanations

Factual Knowledge

1. **a** Weather changes daily, if not more often. Climate follows patterns over long periods of time.

2. **d** Seasons are influenced by the amount of solar radiation reaching a region at a particular time, which in turn is influenced by the tilt of Earth and the movement of Earth around the sun.

3. **e** Warm, moist air rises and cools; dry air descends. This movement, coupled with the rotation of Earth on its axis, contributes to wind patterns.

4. **d** All of the listed factors contribute to desert formation, as does human impact.

5. **b** The Gulf Stream acts as a major warming force for western Europe.

6. **False** Even though biomes are characterized by a dominant species, the distribution of that species varies.

7. **e** Biomes are dynamic entities and are influenced by multiple factors.

8. **a** Climate is responsible for direct exclusion of a species from a particular biome. Species may be indirectly excluded from a biome by other species.

9. **c** Temperature and precipitation affect the distribution of biomes.

10. **b** Terrestrial biomes are classified by their dominant life forms.

11. **False** Though physical characteristics may be similar, species distribution may be quite different.
12. **False** Terrestrial and aquatic biomes have remarkable influences on one another.
13. **e** The changing currents of El Niño have large and widespread effects.
14. **c** biosphere
    b ecosystem
    e habitat
    a ecology
    d biome
15. **b** The ozone layer prevents much of the sun's ultraviolet light from reaching the Earth and thereby protects organisms from mutations.

Conceptual Understanding

1. **d** Many questions have yet to be answered about the environment, its organisms, and the impact of humans on both.
2. **c** The Gulf Stream transfers a tremendous amount of heat energy that moderates the climate of northern Europe; without this influence both summers and winters would be cooler.
3. **b** Terrestrial biomes are categorized according to the dominant vegetation living there.
4. **c** El Niño has caused major, global events.
5. **a** An estuarine environment occurs where salt water and fresh water mingle.
6. **c** These are all characteristics of tundra.
7. **c** The depletion of one or more species by overfishing can dramatically alter the ecological balance between remaining species.
8. **b** Mountains are large geographic barriers that greatly influence climate. For example, mountains may create a rain shadow in which little precipitation falls on the side opposite the prevailing winds.
9. **d** Regions at the equator are tropical because sunlight strikes them directly and as the hot, moist air rises, it cools and causes rain. Large bodies of warm water also influence the climate. Tropical regions are some of the most productive and diverse regions in the biosphere.
10. **d** Humans must live responsibly in order to limit their impact on the biosphere. We have the ability to affect nearly all the factors that influence the biosphere, including climate and the interactions between other organisms and the environment.

CHAPTER 21 Growth of Populations

Getting Started

Below are a few questions to consider prior to reading Chapter 21. These questions will help guide your exploration and assist you in identifying some of the key concepts presented in this chapter.

1. How many days would be required for an *E. coli* population, growing exponentially, to produce as many individuals as there are atoms in the universe?
2. Under what circumstances can exponential growth be observed?
3. What pattern of growth is seen when essential resources become limited?
4. What happened to the reindeer population of Saint Paul Island?
5. How are the population densities of the snowshoe hare and the Canada lynx related?
6. What types of information would biologists need to be able to predict the impact of logging on spotted owl populations?
7. How many humans can the Earth support?
8. In what ways have humans altered their carrying capacity during the past 10,000 years?
9. How many acres are required to support the average citizen of the United States?

A Guide to the Reading

The following concepts typically give students the most difficulty when exploring the content in Chapter 21 for the first time. For each concept, one or more references have been identified that may help you gain a better understanding of these potentially problematic areas.

What Are Populations?

A population is a group of individuals of the same species living and interacting in a common geographic location. As a result of additions to the population through births and immigration and losses from death and emigration, population size has the potential to change with time. Biologists often find it convenient to focus on the biological factors affecting populations. Although the movements of individuals may be locally important, they have no net effect on the total number of individuals; one individual joining a population simply means that it has left another.

For more information on this concept, be sure to focus on:

- Section 21.1, What Are Populations?
- Section 21.2, Changes in Population Size

Patterns of Change

The key to understanding how populations change is to recognize the importance of resources, understand that they have a limited availability, and realize that the individuals of any population require many different resources, each being more or less common. Even when food is abundant, for example, if the appropriate cover required by juveniles is limited, most will be found and taken by predators, and the population will grow slowly or not at all. Populations display a variety of growth patterns in nature, but two patterns are observed repeatedly. Both are best understood as they relate to resource availability. During exponential growth a constant proportion of individuals is added during each unit of time. Such growth occurs only when the population is too small to appreciably impact the quantity of resources provided by the environment. When represented graphically, the number of individuals plotted against time produces a curve that resembles the letter J. Most populations, however, do not maintain a consistent growth rate over time. As resources become limited, the growth rate declines. This type of population growth can be represented by an S-shaped curve. Such populations may appear to be growing exponentially at first, but eventually they stabilize. Biologists use the term *carrying capacity* to describe this level. It is important to understand that these patterns of growth do not represent an either-or situation. Many populations can grow exponentially when their numbers are small, but later convert to an S-shaped pattern as resources become less common. Population growth models can be easily constructed using spreadsheets (see Related Activities), but such representations are often oversimplified. A constant population size implies a constant replenishment of resources. A good example is the regrowth of prairie grasses during the spring from their dormant root systems. Recall, however, that organisms can substantially influence the physical environment. For example, too many hooves seeking a limited amount of grass can result in soil erosion, fewer surviving plants, and less grass for grazing at a later time. Most populations experience variations in carrying capacity, resulting either from direct influences from the population itself or, more commonly, from climatic factors. In the last chapter you read about the El Niño event. You probably realize now that it's the carrying capacity of the open ocean biome that changes when ocean circulation patterns change. When population densities are high, virtually all interactions between population members become more stressful. Competition for food, space, and shelter from predators and weather intensify. Less food, or sunlight, makes an individual more susceptible to disease, and the high number of individuals improves the chances that a disease will quickly spread within the population. When the impact of any factor causes the density of the population to change, that factor can be described as density-dependent. Alternatively, some factors are considered density-independent. One way to distinguish between density-dependence and density-independence is to consider the likelihood of an event relative to population size. Consider predation—when competition for food is low, prey animals can feed in relative safety. When population densities are high, the safe food is quickly eaten, leaving only the more exposed food sources available. Venturing out into the open to nibble a blade of grass may be an animal's last meal.

For more information on this concept, be sure to focus on:

- Section 21.3, Exponential Growth

- In Section 21.4, Growth is limited by essential resources and other environmental factors
- In Section 21.4, Some growth-limiting factors depend on population density; others do not
- Figure 21.11, Same Species, Different Outcomes

The Human Population

Not surprisingly, one of the most important issues facing population ecologists is the future of human growth. Understanding population growth has provided ecologists with some important predictive abilities, and many wonder if these are applicable to the human population. The key to understanding the connection to human growth, and the principles discussed in this chapter, is recognizing that the human carrying capacity has not been static. Technological advances have repeatedly increased the human carrying capacity. Advances in medicine have extended life expectancy and lowered infant mortality. Advances in agriculture have stabilized the food supply in many developed regions. Yet while the human growth curve continues to maintain its J-shape, poverty, starvation, disease, and other indicators of density-dependence are becoming all too common. Will new technologies extend the carrying capacity even further, or do the declining life styles of so many of the world's citizens represent the future of humanity?

For more information on this concept, be sure to focus on:

- Section 21.6, Human Population Growth: Surpassing the Limits?
- Figure 21.12, Rapid Growth of the Human Population

Tying It All Together

Several concepts presented in this chapter build upon concepts presented in previous chapters and may also be revisited and discussed in greater detail in subsequent chapters, including:

What Are Populations?

- Chapter 25 – Section 25.1, Land and Water Transformation

Patterns of Change

- Chapter 22 – Section 22.2, Exploitation
- Chapter 22 – Section 22.3, Competition
- Chapter 23 – Section 23.2, Communities Change over Time

The Human Population

- Chapter 25 – Biology Matters, What's the Size of Your Footprint?

Practice Questions

Factual Knowledge

1. All populations ________.
   a. eventually crash and disappear as a result of environmental deterioration.
   b. contain individuals of one species that interact in the same location.
   c. change their size only in response to the birth of new individuals.
   d. grow exponentially.
   e. all of the above

2. Today, there are on average 30 people per square kilometer living in the United States. This figure is an example of a(n)
   a. carrying capacity.
   b. environmental limit.
   c. population density.
   d. density-dependent limit on growth.
   e. density-independent limit on growth.

3. Two factors that contribute to a decline in population size are _______ and _______.
   a. birth; death
   b. birth; immigration
   c. death; immigration
   d. death; emigration
   e. none of the above

4. In exponential growth, the
   a. numerical increase in the population remains constant for each generation.
   b. population size is usually above the environmental carrying capacity.
   c. population increases by some constant proportion of its current size.
   d. all of the above
   e. none of the above

5. Which of the following can be a limit on population growth?
   a. food
   b. water
   c. space
   d. accumulated wastes
   e. all of the above

6. Refer to Figure 21.6 in your textbook, which shows the growth of a population of *Paramecium caudatum* in a laboratory culture. The straight line of the growth curve, corresponding to the carrying capacity for this population, is the level at which
   a. the number of births equals the number of deaths.
   b. nutrients are extremely abundant.
   c. excess space is abundant.
   d. accumulated wastes are low.
   e. all of the above

7. Which of the following is likely to cause an increase in human population size?
   a. advancements in agriculture
   b. better weather predictions
   c. reducing the number of highway deaths
   d. improvements in medicine
   e. all of the above

8. Volcanic eruptions represent an example of density-dependent limits on population growth. (True or False)

9. The graph shown in Figure 21.8 of your textbook is an example of a
   a. situation in which food and space were not limiting for the population.
   b. population that exceeded its carrying capacity and then crashed.
   c. density-independent limit on population growth.
   d. population cycle.
   e. All of the above

10. Which of the following statements about population growth is *false*?
    a. Tightly linked changes between populations of two species are common.
    b. Different populations of the same species may experience different growth patterns.
    c. Exponential population growth cannot continue indefinitely.
    d. One problem with large populations is that they cause habitat deterioration.
    e. Food shortages, predators, and weather can limit population growth.

11. The spotted owl is an example of a species in which population growth
    a. is declining due to natural factors such as fire and floods.
    b. is currently exponential and numbers are increasing rapidly.
    c. depends on the amount and arrangement of preferred habitat.
    d. has been stable for almost 100 years.
    e. resembles the growth of the human population on Easter Island.

12. If human birth rates were to drop immediately to a level that would not result in population increase, the population would continue to grow for decades. (True or False)

13. One reason human populations have undergone such tremendous growth is that we have
    a. colonized a wide range of new habitats.
    b. greatly increased the carrying capacity of the places where we live.
    c. eliminated predators and many kinds of disease from our lives.
    d. created a situation in which our birth rate exceeds our death rate.
    e. all of the above

14. Sustainable use of resources refers to use that will not deplete critical resources and cause serious damage to the environment. (True or False)

15. Match each term with the best description.
    __ carrying capacity
    __ population size
    __ exponential growth
    __ S-shaped growth curve
    __ population density
    a. one form of rapid population increase
    b. shows common form of population growth
    c. maximum sustainable population size
    d. total number of individuals in a population
    e. number of individuals per unit area

Conceptual Understanding

1. Which of the following examples of population growth in nonhuman organisms shows a pattern similar to that of the humans on Easter Island?
    a. *Opuntia* cactus in Australia
    b. reindeer on Saint Paul Island in Alaska
    c. paramecia in Gause's laboratory
    d. snowshoe hares and lynx in Canada
    e. spotted owls in the U.S. Pacific Northwest
2. Natural populations are easy to define. (True or False)
3. The growth of *Opuntia* cactus in Australia prior to the introduction of the *Cactoblastis* moth was
    a. slow.
    b. moderate.
    c. exponential.
    d. density-dependent.
    e. limited by shortages of nutrients and space.
4. Closed laboratory systems differ from natural systems in having
    a. no immigration or emigration.
    b. limited nutrients.
    c. limited space.
    d. all of the above
    e. none of the above
5. If birth and immigration levels are lower than death and emigration levels, the number of individuals in a population of organisms will
    a. grow.
    b. decline.
    c. stay the same.
    d. double.
    e. none of the above
6. The population growth patterns of giant puffball mushrooms, prickly pear cacti, aphids, and reindeer all demonstrate that
    a. limits exist, and no population can increase in size indefinitely.
    b. every population shows an S-shaped growth curve.
    c. limiting factors on population growth are all density-dependent.
    d. limiting factors on population growth are all density-independent.
    e. natural populations are not sustainable.
7. Many scientists now believe that the human population is at or near its carrying capacity on this planet. (True or False)
8. When carrying capacity drops, the number of individuals that can be supported in a habitat increases. (True or False)
9. Which of the following activities affecting human population growth are considered sustainable?
    a. the use of oil and gas as fuels for heating and transportation
    b. having family sizes greater than two children per female parent
    c. relying more heavily on irrigated crop lands for food production
    d. altering the amount of biodiversity through deforestation
    e. none of the above
10. Even though Americans represent a relatively small percentage of the world's human population, we have a disproportionately large impact on human population growth because we
    a. consume a very high percentage of the world's natural resources.
    b. cannot grow enough food to support ourselves.
    c. don't contribute much to advancements in health and medicine.
    d. have families with very large numbers of children.
    e. have a population doubling time much greater than that of most other countries.
11. If human populations continue to grow at their current rate, by the year 2025 there could be
    a. over 1 billion people on Earth.
    b. over 8 billion people on Earth.
    c. over 1 billion people in the United States.
    d. over 8 billion people in the United States.
    e. fewer people on Earth than there are today.
12. An examination of the numbers of bald eagles in the United States over the past 30 years reveals that
    a. decreasing numbers of prey species prompted declines.
    b. populations can recover from the impact of humans.
    c. species on the road to extinction continue to decline in numbers.
    d. all of the above
    e. none of the above

Related Activities

- Consider the 1993 hantavirus outbreak in the American Southwest. This is a good example of how a deadly human disease is affected by population growth in nonhuman organisms (in this case, deer mice). Use your library or the World Wide Web to research human diseases that are spread by animals such as mosquitoes, ticks, and rodents. Then write a short essay on the similarities and differences between these diseases. In particular, focus on how the population growth of the animals carrying the diseases influences the likelihood of human infection.
- Study the Web site maintained by the organization Zero Population Growth (www.zpg.org) and write a short essay describing the organization's goals and activities. Include your own opinion of the political and social agendas of this organization.
- Write a letter to the editor of your local newspaper on the issue of human population growth. Briefly describe the biology behind the problem, as well as your own personal views about how we can minimize the potential for ecological disaster. Be sure to include suggestions for ways that we can help reduce the problem of overpopulation.
- Use Excel to examine J-shaped and S-shaped growth. Create a time series in Column A. Enter 1 in cell A1. Enter this formula in Cell A2 = (A1+1), and use the fill handle to fill down to A35. Enter identical initial population sizes (20) in cells B1 and C1, an initial lambda (1.25) in cell D1, and a carrying capacity value in cell E1 (350). To model J-shaped growth, enter this formula in cell B2 = B1+B1*(C1-1), and use the fill handle to fill down through B35. To model S-shaped growth, enter this formula in cell C2 = C1+C1*((D1-1)*(E1-C1)/E1), and fill down through C35. Select the block A1:C35 and use the Chart Wizard to construct a Scatter Graph with smoothed lines. To see S-shaped growth on the graph, you will need to adjust the maximum value for the Y axis. Simply change the initial population size, lambda, or carrying capacity to explore the impact these values have on the growth curves. Write a brief summary describing the difference between the growth models.

Answers and Explanations

Factual Knowledge

1. **b** This is the basic definition of any population.
2. **c** Population measurements that indicate the number of individuals per unit area are known as densities.
3. **d** Death and emigration are ways that individuals leave a population, and so they contribute to a decline in numbers.
4. **c** Exponential growth occurs when the population increases by some proportional amount (doubling, for example) in each successive generation. Such growth produces a very rapid increase in numbers.
5. **e** The first three factors represent important resources that populations need in order to reproduce. The fourth is a potential negative factor for survival and reproduction. Because reproduction is an important contributor to population growth, all four factors can potentially limit growth.
6. **a** At this point there is no net population growth, and so by definition births are equal to deaths. The other three conditions are more likely to occur when the population size is well below the carrying capacity of the environment.
7. **e** All of these improvements in human life would likely increase survival and reproduction and therefore would lead to increases in population size.
8. **False** Volcanic eruptions fall into the same category as fires, floods, and weather variations. Such phenomena are density-independent limits on growth.
9. **b** Clearly, the population of reindeer grew well initially, as evidenced by the exponential rise in numbers during their first 20 years on Saint Paul Island. However, at its peak the population exceeded the carrying capacity of the environment, and then it crashed to extinction before the environment had a chance to recover enough to sustain the population.
10. **a** Long-term, tightly linked changes between two populations, as seen with the snowshoe hares and the lynx, are actually rather uncommon in nature.
11. **c** Growth of spotted owl populations appears to depend on the number and location of old growth forests, the species' preferred habitat.
12. **True** If human birth rates dropped immediately to a level that would allow replacement but not increase, the population would continue to grow for at least 60 years because of the huge number of children who have not yet reproduced.
13. **e** All of these factors increase survivability in one way or another. The end result is more births and a lower death rate, both of which contribute to population increases.
14. **True** Sustainable use of resources does not deplete the resources or seriously damage the environment. In the past, humans have used resources in a nonsustainable manner.

15. **c** carrying capacity
    d population size
    a exponential growth
    b S-shaped growth curve
    e population density

Conceptual Understanding

1. **b** Both populations showed exponential growth, exceeded their carrying capacities, and then crashed.
2. **False** Natural populations are often difficult to define, and the definition may depend on the organism and particular research question.
3. **c** An increase from zero to a coverage of over 243,000 square kilometers in about 90 years easily qualifies as exponential growth!
4. **a** Unlike natural systems, closed laboratory systems have no immigration or emigration. Space and nutrients are usually limited in both closed and natural systems.
5. **b** If the factors that remove individuals from the population exceed those that add individuals, the population will decline.
6. **a** This is an environmental "fact of life" for all organisms in all habitats.
7. **True** Although it is impossible to predict accurately the maximum carrying capacity for a highly adaptable species such as humans, most experts agree that we are pushing the limits of Earth's ability to support our growing population.
8. **False** A decrease in carrying capacity means that the habitat can support fewer individuals.
9. **e** All of these practices lead to some form of environmental degradation, which is not a sustainable strategy for maintaining a healthy human population.
10. **a** Americans are enormous consumers of energy and natural resources, far in excess of our percentage of representation in the world's human population.
11. **b** 8.3 billion, to be more precise!
12. **b** Bald eagles are an example of a species that recovered from human interference. Several decades ago, our use of DDT in pesticides caused reproductive failure in eagles and other predatory birds. Banning DDT has led to the substantial recovery of their populations.

CHAPTER 22

Interactions among Organisms

Getting Started

Below are a few questions to consider prior to reading Chapter 22. These questions will help guide your exploration and assist you in identifying some of the key concepts presented in this chapter.

1. What term do ecologists use to describe the interaction that benefits both species?
2. How would extinction of the yucca plant affect the yucca moth?
3. What are the three groups of consumers that exploit other species for food?
4. Why are the spines on cacti that have been grazed longer than those on the plants that have not?
5. Why are many of the frogs found in the tropics so brightly colored?
6. What behavioral strategy do pigeons use to reduce goshawk predation?
7. Why was the American chestnut virtually defenseless against the fungus that causes chestnut blight?
8. Under what circumstances will character displacement occur?

A Guide to the Reading

The following concepts typically give students the most difficulty when exploring the content in Chapter 22 for the first time. For each concept, one or more references have been identified that may help you gain a better understanding of these potentially problematic areas.

Mutualism

Interactions among organisms affect individuals, populations, communities, and ecosystems. Ecologists classify these interactions by whether the interaction is beneficial or harmful to each of the interacting species. The key to understanding this classification is understanding that ecologists use the terms *beneficial* and *harmful* in the context of reproductive success. Any change that improves an organism's chances of mating or extends its lifespan would be considered beneficial. The death of an organism before it reproduces is, of course, the ultimate harm. Many varieties of mutualism occur. One of the most common is the bacteria that inhabit the digestive systems of animals and assist in the digestive process. As you read about other examples of mutualism, such as seed dispersal and pollination, evaluate each for the principles discussed above. Several misconceptions about mutualism are common. You might think, for example, that the relationship benefits both species equally, but this is generally not correct. It is common for each participant to experience some cost because of the relationship. Mutualism simply occurs when the benefits are greater than the costs. Some mutualistic relationships have become surprisingly specialized. In the most extreme examples, each participant is entirely dependent on the other, and neither can use an otherwise environmentally suitable habitat unless the other is present. As a result, mutualisms can influence the distribution and abundance of species. Similarly, should one species become locally or globally extinct, the other species will experience the same fate.

For more information on this concept, be sure to focus on:

- In Section 22.1, There are many types of mutualisms
- In Section 22.1, Mutualists are in it for themselves
- In Section 22.1, Mutualisms can determine the distribution and abundance of species

Exploitation

Exploitation includes those interactions where one species benefits at the expense of another. Virtually all consumers, defined as organisms that are unable to produce their own food, are dependent on exploitation, making it one of the most common interactions in nature. Even those consumers that appear to be feeding on organic debris will be consuming the bacteria that first colonized the item. Natural selection has equipped the majority of prey species with a variety of defenses that work to reduce the harmful effects of exploitation. Induced defenses are those whose activity is stimulated by an attack. Plants often produce toxic chemicals only after grazing begins. Consider the elaborate response of the human immune system to infection by pathogenic bacteria or parasites. Prey defenses create an opportunity for changes within the consumer. For those species with long-term associations, many cycles of adaptation may have taken place, resulting in some unusual specializations. Consider the rough-skinned newt. Its skin glands produce a toxin so potent it is considered one of the most poisonous organisms on Earth. Only one predator has evolved the ability to overcome this toxin, the common garter snake. High toxicity is a personally marginal defense if the individual must die for it to be useful. The majority of toxic prey species advertise their toxicity through warning colorations. Bright, bold, or otherwise, highly visible patterns serve to communicate to potential consumers that the prey item will be distasteful or harmful. A variety of additional strategies have evolved within prey populations that do not involve toxicity. Consider the herding, flocking, or schooling behavior of some animals, birds, and small fish. Despite the fact that a group is more easily located than a single individual, the chances of any individual surviving an attack are much greater when within the group. The development of defensive and offensive strategies by natural selection is slow, often requiring dozens or hundreds of generations. The introduction of prey or consumer species into an ecosystem is often accompanied by extensive disruption. You read earlier about the introduction of the zebra mussel, a prey species growing without the control exerted by its natural predators. This chapter describes the effect of chestnut blight, a fungal consumer introduced to a prey population, the American chestnut, which lacked any natural defenses. As has been true in many other similar situations, the American chestnut is now virtually extinct where its distribution overlaps that of the fungus.

For more information on this concept, be sure to focus on:

- Section 22.2, Exploitation
- Figure 22.8, Safety in Numbers

Competition

Competition is considered to be an interaction where both participants are harmed. There are two main types of competition. In interference competition one species is prevented from using a resource by the actions of the competitor species. The decline of coyotes following the reintroduction of wolves in Yellowstone National Park provides a good example. Although not eating the same foods as coyotes, wolves are highly territorial. Their territories left few spaces where coyotes could hunt without the risk of attack from wolves. Although any given wolf territory contains an abundance of unused food items suitable for the coyote, the action of the wolves interferes with the coyote feeding, and the population has declined. In exploitation competition the resource is used by both species, and its use by one competitor prevents its use by the other. Competition for space is one of the most common examples of exploitation competition. On rocky intertidal shorelines throughout the world, competition occurs for attachment sites. Ecologists had long wondered why so many intertidal organisms lived in horizontal bands with clear boundaries between them. Research has shown that the lower limit of the distribution of one species is almost always the result of a biological interaction with the species below, typically the better competitor.

For more information on this concept, be sure to focus on:

- In Section 22.3, Competition can limit the distribution and abundance of species
- Figure 22.10, What Keeps Them Apart?

Tying It All Together

Several concepts presented in this chapter build upon concepts presented in previous chapters and may also be revisited and discussed in greater detail in subsequent chapters, including:

Mutualism

- Chapter 3 – in Section 3.5, Some fungi live in beneficial associations with other species

Exploitation

- Chapter 16 – Section 16.2, Mechanisms of Evolution
- Chapter 24 – Section 24.4, Energy Capture in Ecosystems

Competition

- Chapter 21 – Section 21.4, Limits to Population Growth
- Chapter 23 – in Section 23.1, Keystone species have profound effects on communities

Practice Questions

Factual Knowledge

1. Pollinator mutualisms are special interactions involving ________, which receive food or a place to lay eggs, and ________, which receive pollen from others of their kind.
   a. insects; plants
   b. plants; insects
   c. parasites; plants
   d. predators; plants
   e. none of the above

2. Mutualists are altruistic in that they react to the other species in ways that do not benefit themselves. (True or False)

3. In a mutualistic relationship, only the species directly involved benefit. (True or False)

4. Warning coloration is a result of
   a. consumers placing selective pressure on a species.
   b. victims evolving faster than consumers.
   c. behavior changing the physiology of an organism.
   d. all of the above
   e. none of the above

5. Consumers can alter which of the following characteristics of victims?
   a. distribution
   b. abundance
   c. behavior
   d. all of the above
   e. none of the above

6. Group interaction is often an alteration of behavior to
   a. reduce selective pressures.
   b. avoid consumption by another species.
   c. speed character displacement.
   d. all of the above
   e. none of the above

7. Species compete indirectly for shared resources in ________ competition.
   a. interference
   b. exploitative
   c. interrelation
   d. infringement
   e. none of the above

8. The level of competition between species depends on
   a. availability of resources.
   b. population density.
   c. group interaction of organisms.
   d. all of the above
   e. none of the above

9. Two species of Galápagos finches have very similar beak sizes on islands where only one species is found, but different beak sizes on islands where they are found together. This is an example of
   a. convergent evolution.
   b. character displacement.
   c. competitive fitness.
   d. ephemeral differences.
   e. all of the above

10. Interactions among organisms can have effects at which of the following levels?
    a. individual
    b. population
    c. community
    d. ecosystem
    e. all of the above

11. The human relationship to plants and animals used for food can best be described as a ________ relationship.
    a. consumer–victim
    b. mutualistic
    c. competitive
    d. both a and b
    e. both a and c

12. Interactions in which the consumer lives within the host and does slow damage to the host are referred to as
    a. commensalism.
    b. parasitism.
    c. mutualism.
    d. competition.
    e. none of the above

13. Extinction can occur through competition. (True or False)

14. Which of the following is a symbiotic relationship?
    a. gut inhabitant mutualism
    b. parasitism
    c. intraspecific competition
    d. both a and b
    e. none of the above

15. Match each term with the best description.
    __ behavioral mutualism
    __ character displacement
    __ interference competition
    __ parasite
    __ predator
    a. forms of competing species evolve differently over time to lessen competition
    b. the member of a consumer–victim relationship that hunts and kills prey

c. a species alters its behavior to benefit another in order to gain benefits for itself
d. competition through preventing access to resources
e. the member of a consumer–victim relationship that attains its resources by living off of a host

Conceptual Understanding

1. Two organisms share the same habitat and consume the same food source. Neither prevents the other from obtaining the food source. This is an example of ________ competition.
   a. interference
   b. exploitative
   c. intimate
   d. all of the above
   e. none of above

2. In lichens, fungus and blue-green algae live together without causing significant harm to one another. Attempts to grow either the algae or the fungi independently result in the death of the organism. This interaction is an example of
   a. parasitism.
   b. mutualism.
   c. competition.
   d. all of the above
   e. none of the above

3. Gazelles in a group have a 75 percent chance of escaping hungry lions. Solitary gazelles have only an 8 percent chance of escaping. In this case, ________ behavior is advantageous for avoiding predation.
   a. mutualistic
   b. individual
   c. group
   d. all of the above
   e. none of the above

4. Barnacles on the Scottish shore, as shown in Figure 22.10, illustrate ________ competition.
   a. interference
   b. exploitative
   c. intimate
   d. all of the above
   e. none of the above

5. Lizards on a chain of islands prefer to eat the berries of the Make-Believe bush. Two species of lizards compete for this resource. It is noticed that on Fantasy Island the two species of lizards differ in the length of their legs, with one species grazing on berries at the bottom of the bush and the other species grazing on the top of the bush. However, on the other two islands of the chain, where only one or the other species is present, they appear to have legs of the same length. This is an example of
   a. character development.
   b. character displacement.
   c. character morphogenesis.
   d. all of the above
   e. none of the above

6. Your friend Sue just returned from a trip abroad with symptoms of severe diarrhea. Cultures show that she has an intestinal parasite. Your friend Sue has just become part of a ________ interaction.
   a. mutualist
   b. predator–prey
   c. consumer–victim
   d. competitive
   e. all of the above

7. Cows eating grass represent what type of consumer–victim interaction?
   a. parasitism
   b. predation
   c. herbivory
   d. infection
   e. all of the above

8. If a new predator is introduced to an area with a type of prey that is limited in number, the prey species is likely to
   a. decrease, and possibly become extinct.
   b. increase.
   c. remain stable.
   d. initially increase, then decrease.
   e. none of the above

9. Human interference has caused ecological imbalances by
   a. affecting consumer–victim interactions.
   b. altering competition between species.
   c. eradicating common parasites.
   d. all of the above
   e. none of the above

10. The diversity and abundance of species are altered by
    a. competition.
    b. consumer–victim interactions.
    c. mutualisms.
    d. all of the above
    e. none of the above

11. Mutualistic relationships
    a. do not have costs.
    b. evolve when the benefits of the interaction outweigh the costs.
    c. are rare.
    d. are limited to plants and animals.
    e. none of the above

12. Which of the following statements about parasites is *true*?
    a. They often harm but don't immediately kill their hosts.
    b. They can alter the behavior of their hosts.
    c. They may kill their host if this promotes transmission to another host.
    d. all of the above
    e. none of the above

Related Activities

- Make a list of 10 common species interactions in a habitat near your school. Next to each one, identify which type of interaction it is and the role that each member plays in the interaction.
- For each type of interaction discussed in this chapter, think of an example that involves humans. Briefly describe the interaction and the other organisms involved.
- Using your library or the World Wide Web, find an example of a situation in which competition or consumption has led to the extinction of a species. Write a one-page essay describing the example.
- Using the library or the World Wide Web, investigate how prey species can benefit from living in groups. Then examine whether living in groups makes some animals more susceptible to predation. Outline your findings in a short essay.

Answers and Explanations

Factual Knowledge

1. **a** In pollinator mutualisms, insects pollinate plants in exchange for food or other rewards.
2. **False** Mutualists receive benefits for themselves. Without these benefits, the interaction would not exist.
3. **False** In addition to the species directly involved in a mutualistic relationship, other organisms may also benefit. For example, coral reefs result from the mutualistic relationship between corals and photosynthetic algae, and these reefs provide a home to many other organisms.
4. **a** Those with warning coloration are less likely to get eaten, and their genes are passed on.
5. **d** Consumers can alter the distribution, abundance, and behavior of their victims.
6. **b** Groups have better detection and warning mechanisms than individuals do.
7. **b** The resource is available to both competitors.
8. **d** Limitation of resources and higher population density will result in greater competition among organisms. Group interactions may limit or increase competition depending on the circumstances.
9. **b** Character displacement is the altering of a character in areas of high competition so that other resources may be exploited.
10. **e** All these levels are influenced by organismal interactions.
11. **d** It is a consumer–victim relationship in that humans kill and eat the animals and plants. It is also mutualistic in that, through agriculture, many more of those plants and animals are produced than under natural circumstances.
12. **b** This is the classic definition of a parasitic interaction.
13. **True** Competitive pressure may lead to extinction.
14. **d** Gut inhabitant mutualism and parasitism are symbiotic relationships because in each relationship two different species live in direct contact.
15. **c** behavioral mutualism
    a character displacement
    d interference competition
    e parasite
    b predator

Conceptual Understanding

1. **b** In exploitative competition, competing species all have access to the resource.
2. **b** Lichens exemplify mutualism.
3. **c** Predation is a driving force behind group behavior in this case.
4. **a** One group of barnacles interferes with the other's access to resources.
5. **b** These similar species show divergence of characters as a result of competition.
6. **c** She and her parasite are in a consumer–victim interaction.
7. **c** The interaction is herbivory.
8. **a** The initial result will be a reduction in the number of prey species. The prey may well become extinct, as happened to many species of birds in New Zealand following the introduction of house cats.

9. **d** There are very few realms in which humans have not had an impact.
10. **d** All types of interactions between organisms affect diversity and abundance.
11. **b** Mutualistic relationships are usually not cost-free; instead, such relationships evolve when the benefits exceed the costs for the species involved. In addition, mutualisms are very common and are not limited to plants and animals.
12. **d** Parasites harm their hosts and may alter their behavior. Some parasites kill their host if such killing promotes transmission to the next host in their life cycle.

CHAPTER 23 Communities of Organisms

Getting Started

Below are a few questions to consider prior to reading Chapter 23. These questions will help guide your exploration and assist you in identifying some of the key concepts presented in this chapter.

1. In what way is the digestive system of a deer like a temperate coniferous forest?
2. How does a food web differ from a food chain?
3. What actions by the common Pacific seastar, *Pisaster ochraceus*, cause ecologists to consider it a keystone species?
4. What distinguishes a climax community?
5. What feature, if intact, permits secondary succession to begin following a disturbance?
6. Queensland, Australia, is located near the equator; what was its location 1 billion years ago?
7. In what ways does eutrophication alter the appearance and habitat potential of a lake?
8. What happened after the comb jelly was accidentally introduced to the Black Sea?
9. Which New Mexico habitat will recover most quickly, a grassland used for cattle grazing or one used to test nuclear bombs?

A Guide to the Reading

The following concepts typically give students the most difficulty when exploring the content in Chapter 23 for the first time. For each concept, one or more references have been identified that may help you gain a better understanding of these potentially problematic areas.

The Effects of Species Interactions on Communities

Communities are composed of populations of different species in a particular environment. That environment may be as large as a forest or as small as the digestive system of a deer. Communities are described and understood not by their size, but on the basis of the interactions that occur between those species. The interactions within a community are complex and subject to the effects of many factors, both natural and human-caused. The most thoroughly studied aspect of communities is the pathway of energy and nutrient distribution. Recall that all organisms can be classified as producers or consumers. Consumers obtain their energy and nutrients by eating all or portions of other organisms or their remains. Only one type of producer is recognized. However, ecologists distinguish between consumers on the basis of the type of food they eat. Primary consumers obtain the majority of their energy directly from producers. Secondary consumers feed primarily on primary consumers, while tertiary consumers feed primarily on secondary consumers. It is important to understand that these terms reflect concepts that help us understand nature, rather than mandatory categories. For example, when a deer grazes on woodland shrubs, it consumes not only the leaf but also bacteria and fungal spores. Although small in comparison to the energy in the leaf, the bacteria and fungi, which are both consumers, represent a portion of the deer's energy budget. As you read more about communities, remember that nature is much more complex and intriguing than our simple descriptions sometimes suggest. Communities are often disproportionately affected by one or a few species called *keystone species*. If removed, the entire composition of the community may change. The experiments with seastars described in the text illustrate this concept. Once seastars were removed from the community, mussels were able to outcompete virtually every other attached intertidal organism. Such results were not entirely predictable. Seastars were neither a dominant nor an abundant community member. The loss of even a single species through natural or human-influenced activities can have disastrous effects on community structure.

For more information on this concept, be sure to focus on:

- In Section 23.1, Food webs consist of multiple food chains
- Figure 23.3, A Food Web
- In Section 23.1, Keystone species have profound effects on communities
- Figure 23.4, A Keystone Species

Communities Change over Time

All communities change over time. One source of change involves biological interactions and is called *succession*. In succession, a series of communities occupies a site in a somewhat orderly and predictable fashion. These living organisms can alter the physical environment. An existing plant community exerts powerful effects on the structure of the soil. Its roots penetrate the soil; its leaves and roots may decompose and add organic material, and the shade it creates may reduce evaporation and raise moisture levels. With time, these improvements may help competitors displace some species, and the composition of the community will change. Ecologists distinguish between primary succession, which usually begins on a newly created habitat, and secondary succession, which occurs after a disturbance removes the existing community but leaves intact the soil and the roots and seeds it contains. Change in communities may be unrelated to biological processes. Climate is one of the most influential factors on community and biome development. The Earth's climate regularly cycles through long-scale periods of cooling and warming. We are presently in a relatively warm interglacial period. Should the climate cycle as it has for the past 400,000 years, we can anticipate another cooling trend and ice age in the near future. Even if climatic cycles were to cease, habitats would continue to experience slow climatic change because of continental drift (Chapter 19). The slow movement of the continental land masses eventually moves them to different latitudes and the climates associated with them. A slow northward drift has moved North America during the past 135 million years from a semitropical location to its current temperate and arctic position.

For further information on this concept, be sure to focus on:

- In Section 23.2, Succession establishes new communities and replaces disturbed communities
- Figure 23.5, Succession
- In Section 23.2, Communities change as climate changes
- Figure 23.7, Climate Change, Community Change

Recovering from Disturbance

One of the most important questions facing ecologists involves the potential for the recovery of a community after disturbance. Natural disturbances, such as floods, volcanic eruptions, fires, and windstorms, have always challenged communities. Not surprisingly, most communities are able to recover from such disruptions. More alarming, however, is the relatively sudden emergence of human-induced disturbance. Lacking the long-term evolutionary experience that characterizes the relationship between communities and natural disturbance, most communities are less resilient when exposed to human impacts. Fortunately, some disturbances are recoverable. The text describes the consequences of nutrient enrichment, a process called *eutrophication*, in aquatic biomes. Once the source of the nutrient is identified and stopped, most aquatic systems recover. Introduced species represent a much more serious challenge. In some instances, the additional introduction of a consumer may curtail the ecological damage caused by the initial introduction. The control of both the comb jelly and the prickly pear cactus reflects such an outcome. Despite these two successsful examples, introduced species do not typically result in happy endings. Many species have been driven to extinction by introduced species, and once these species are gone, the community will never return to its original condition. Certain agricultural practices have the potential to degrade communities in ways that may also permanently alter them. Throughout the world, overgrazing is converting grasslands into deserts. Once the grassland soil structure is lost, it may take thousands of years to recover. If its initial existence depended on interactions with particular species, it may never become reestablished. Since humans depend on the continuous functioning of communities, these issues take on serious implications. A variety of moral, ethical, and practical considerations confront us as we consider our impact on the environment.

For further information on this concept, be sure to focus on:

- Section 23.3, Recovery from Disturbances
- Figure 23.9, Changes in the Black Sea
- Figure 23.11, Overgrazing Can Convert Grasslands into Deserts

Tying It All Together

Several concepts presented in this chapter build upon concepts presented in previous chapters and may also be revisited and discussed in greater detail in subsequent chapters, including:

The Effects of Species Interactions on Communities

- Chapter 22 – Section 22.2, Exploitation

Communities Change over Time

- Chapter 20 – Section 20.4, Life on Land
- Chapter 20 – Section 20.5, Life in Water

- Chapter 25 – Section 25.3, Changes in the Global Nitrogen Cycle
- Chapter 25 – Section 25.5, Global Warming

Recovering from Disturbance

- Chapter 25 – Biology Matters, What's the Size of Your Footprint?
- In Interlude E, Sources of Hope for the Future

Practice Questions

Factual Knowledge

1. Artificial reefs placed in the oceans are unique biological opportunities for observing
   a. secondary succession.
   b. primary succession.
   c. climax communities.
   d. habitat destruction.
   e. none of the above

2. Human impact can be devastating on island communities because
   a. native species are not equipped to deal with introduced species.
   b. humans may unintentionally introduce species.
   c. introduced species often have no competition or consumers.
   d. all of the above
   e. none of the above

3. Which of the following best describes a community?
   a. populations of different species that live in the same area
   b. a single species and its habitat
   c. a single population and its habitat
   d. all of the above
   e. none of the above

4. Food chains differ from food webs in that
   a. food chains are a single sequence of who eats whom in a community.
   b. food webs represent the complex interactions among food chains.
   c. food chains better represent the entire community.
   d. both a and b
   e. both a and c

5. A cow eating grass is an example of a
   a. secondary consumer.
   b. primary consumer.
   c. producer.
   d. tertiary consumer.
   e. none of the above

6. Terrestrial producers are mainly
   a. green plants.
   b. zooplankton.
   c. herbivores.
   d. carnivores.
   e. none of the above

7. A community's health depends on
   a. human intervention.
   b. species interaction.
   c. a specific ratio of producers to consumers.
   d. all of the above
   e. none of the above

8. In a climax community,
   a. species are replaced cyclically.
   b. species are not replaced.
   c. new species are evolving.
   d. all of the above
   e. none of the above

9. Recovery from disturbance within a community occurs
   a. randomly.
   b. within a defined time span.
   c. variably, depending on the disturbance and the conditions.
   d. all of the above
   e. none of the above

10. The greater the number of consumer levels in a food chain, the further away from the producer the final consumer is. For example, a hawk eating a fish that ate an insect that ate grass is a tertiary consumer. (True or False)

11. Secondary succession refers to the second phase of succession that occurs after a disaster. (True or False)

12. Overgrazing, overlogging, overburning, and overfarming are all examples of human impact on
    a. communities.
    b. succession.
    c. consumers.
    d. all of the above
    e. none of the above

13. Which of the following factors influence communities?
    a. climate
    b. species interactions
    c. feeding relationships among organisms
    d. succession
    e. all of the above

14. Match each term with the best description.
    __ primary succession
    __ keystone species
    __ climax community
    __ consumer
    __ producer
    a. organisms that obtain nutrients from the sun
    b. the colonization of a new environment
    c. species necessary for keeping other species in check
    d. community that has completed succession
    e. organisms that consume other organisms

Conceptual Understanding

1. During the last century, a volcanic eruption destroyed an entire island. Several years later a new island, sterile and devoid of life, was born on the spot where the original island had disappeared. Scientists were able to view the island before colonization began and as it established itself. These scientists were witnessing
    a. primary succession.
    b. secondary succession.
    c. foundation building.
    d. keystone placement.
    e. none of the above

2. Floods ravaged much of the North Carolina coast in the wake of Hurricane Floyd in 1999. Many communities that were destroyed are in the process of rebuilding. These communities are undergoing
    a. primary succession.
    b. secondary succession.
    c. foundation building.
    d. keystone placement.
    e. none of the above

3. A biology laboratory kept a saltwater tank in which to display aquatic organisms. The person in charge of the tank decided to trade the algae-eating snails for some brightly colored fish. Within weeks, the tank was coated with red algae and the fish began to die. Within a month, little was left living in the tank. In removing the snails, the person had removed the ________ species.
    a. foundation
    b. producer
    c. keystone
    d. consumer
    e. none of the above

4. You have a pet snake that is fed rats. The rats eat a grain-based diet. The snake is a
    a. primary consumer.
    b. producer.
    c. secondary consumer.
    d. herbivore.
    e. none of the above

5. In a pond, you notice a snapping turtle, tadpoles, algae, various species of aquatic insects, and many fish. This pond community represents a food
    a. chain.
    b. web.
    c. pyramid.
    d. all of the above
    e. none of the above

6. Levees along the Mississippi River have occasionally turned annual minor spring floods into floods of devastating power. This is an example of
    a. human impact on a community.
    b. loss of cyclical patterns of change in a community.
    c. community destruction after a major flooding event.
    d. all of the above
    e. none of the above

7. Succession in a community
    a. always occurs at a stable rate.
    b. always reaches the climax stage.
    c. may never be complete.
    d. all of the above
    e. none of the above

8. Changes in the makeup of communities in Australia result from
    a. succession.
    b. climate change.
    c. human impact.
    d. all of the above
    e. none of the above

9. A cow in a pasture is an example of
    a. a food chain within a community.
    b. how communities may exist within communities.
    c. how human needs affect communities.
    d. all of the above
    e. none of the above

10. Changes caused by humans to communities
    a. raise the economic value of the communities.
    b. enhance the aesthetic value of most communities.
    c. occur rapidly across large areas.
    d. all of the above
    e. none of the above

11. Keystone species are always predators. (True or False)

12. Introduced species have dramatic effects on island communities because
    a. species living on the islands have evolved in isolation.
    b. their predators are absent from islands.
    c. their competitors are absent from the islands.
    d. all of the above
    e. none of the above

Related Activities

- Find an example of succession in your community. Discuss the changes that have occurred, the species that are now present, and the type of succession you are observing.
- Develop a brochure that highlights what your community is doing to minimize the impact humans are having on the environment.
- Investigate your state's land preservation policies. What measures are being taken to set aside lands for preservation, and what has been the impact of these measures on the communities of organisms residing in the protected areas?

Answers and Explanations

Factual Knowledge

1. **b** This is an environment ready to be newly colonized, and it provides a perfect opportunity to observe primary succession.
2. **d** Introduced species have been devastating Hawaii as well as Australia and New Zealand.
3. **a** A community consists of a group of interrelating species and the area in which they live.
4. **d** The complex relationships among food chains are best described using a food web.
5. **b** The grass is the producer, and the cow is the primary consumer.
6. **a** Plants are the main terrestrial producers.
7. **b** Interaction among species in the community keeps the community healthy.
8. **b** A climax community is a stable community with no new species replacing existing ones.
9. **c** Recovery of a community from disturbance is extremely variable and depends on a number of factors.
10. **True** Consumers eat other organisms. Primary consumers are herbivores, secondary consumers are predators that eat herbivores, and tertiary consumers eat secondary consumers.
11. **False** Secondary succession refers to the recolonization of an area after a disturbance.
12. **a** Humans are responsible for some, if not all, of the largest impacts on communities.
13. **e** Communities are influenced by interactions among species and by food webs and chains. In addition, communities change as a result of succession and climate change.
14. **b** primary succession
    c keystone species
    d climax community
    e consumer
    a producer

Conceptual Understanding

1. **a** A barren, uncolonized island is a textbook example of primary succession.
2. **b** Recolonization after a disturbance is referred to as secondary succession.
3. **c** The keystone species is the species in a community that keeps the others in check.
4. **c** In this example, the rat is a primary consumer that eats the producer, and the snake is a secondary consumer because it eats the rat.
5. **b** This is definitely a food web with many food chains represented.
6. **d** Since the Mississippi's flooding in 1993, many ecologists and city planners have begun to discuss how to reduce the impact of floods on both humans and the environment.
7. **c** Climax communities (the end point of succession) are very unusual.
8. **d** All of the factors listed contribute to community change.
9. **d** The cow is in the pasture because humans need it for food. It is a primary consumer of grass. It contains a community within itself. It is a very important cow!
10. **c** Change caused by humans to communities tends to occur rapidly and across large areas. Such change usually decreases the economic and aesthetic value of the communities.

11. **False** Although some keystone species are predators, others are producers, herbivores, or pathogens.

12. **d** Species introduced to islands often increase in number dramatically because their predators and competitors are absent. In addition, because species that normally live on islands often evolve in isolation, they are ill-equipped to cope with introduced species.

CHAPTER 24 Ecosystems

Getting Started

Below are a few questions to consider prior to reading Chapter 24. These questions will help guide your exploration and assist you in identifying some of the key concepts presented in this chapter.

1. How did New York City decide to correct its problem with poor water quality?
2. Why are hawks and owls so much less common than the rodents they eat?
3. How is agriculture in the Midwest related to the dead zone in the Gulf of Mexico?
4. If energy can neither be made nor destroyed, why can't it be recycled in an ecosystem?
5. What two factors best predict the level of net primary productivity for any region?
6. In what way does the cycling of phosphorus differ from the cycling of all the other major nutrients?
7. What pollutant is responsible for most acid rain?
8. How successful were the scientists that built Biosphere II in designing an artificial ecosystem?

A Guide to the Reading

The following concepts typically give students the most difficulty when exploring the content in Chapter 24 for the first time. For each concept, one or more references have been identified that may help you gain a better understanding of these potentially problematic areas.

Ecosystems

An ecosystem consists of a community of organisms along with the physical environment in which those organisms live. Two primary activities characterize an ecosystem, the unidirectional flow of energy and the recycling of nutrients. Ecosystems often have no physical boundaries and vary in size from areas as small as mud puddles to entire oceans. With no specific size requirement and no discernable boundaries, it might seem impossible to identify an ecosystem. Rather than using physical features, ecologists delineate ecosystems using functional criteria. All the organisms that utilize the energy and nutrients from a single set of producers belong to the ecosystem. Organisms obtaining their energy and nutrients elsewhere belong to a different ecosystem. Of course, such a simple explanation is rarely suited to the complexities of nature. Many organisms are highly mobile (the Arctic tern, for example, flies 22,000 miles annually) and frequently move between ecosystems.

For more information on this concept, be sure to focus on:

- Section 24.1, How Ecosystems Function: An Overview
- Figure 24.1, How Ecosystems Work

Energy Capture in Ecosystems

All living organisms require energy. Consumers obtain energy by digesting the complex chemical compounds they obtain from their food. Producers, however, collect their energy from nonliving sources, primarily the sun. Producers use solar energy, carbon dioxide, and water (Chapter 8) to photosynthesize the energy-rich compounds that will eventually nourish their bodies, primary consumers directly, and all other consumers indirectly. Because photosynthesis is the essential first step in ecosystem function, ecologists find it useful to characterize different ecosystems by the amount of photosynthesis that takes place within them. Total photosynthesis, however, would be a misleading measurement, since a portion of the captured energy is used by the producer. Subtracting this quantity from total photosynthesis gives a value called net primary productivity (NPP). It is the standard measure of ecosystem productivity and has been used worldwide to compare one ecosystem with another. Patterns of NPP are predictably variable. For example, since the intensity of sunlight varies with the latitudes, you might predict high NPP near the equator, with declines as latitude increases. Of course, photosynthesis also requires water. The Earth's atmospheric convection cells produce complex patterns of rainfall. Areas receiving ample sunlight and abundant rain, like the equator, do experience high NPP. Areas with abundant sunlight and little rain have a low NPP. The NPP in aquatic biomes is less predictable. Aquatic organisms are able to use much lower light intensities than terrestrial plants since light penetrates the water column poorly. Productivity in aquatic biomes is often more dependent on the availability of dissolved nutrients. Rivers and streams typically deliver these, creating highly productive wetlands such as estuaries and marshes where the nutrients begin to accumulate. Human activity can significantly increase or decrease NPP. Nutrient enrichment from agricultural runoff or sewage outflows can produce almost unimaginable increases in productivity in aquatic ecosystems. Growth can be so abundant that the water becomes opaque from the organisms it contains. Recall that exponential growth seldom lasts for long. These organisms die rapidly, followed by bacterial decomposition that can entirely deplete the dissolved oxygen within the area. Oxygen depletion has now become an apparently permanent feature of the Gulf of Mexico. Such massive disruptions to ecosystem function are almost always detrimental.

For more information on this concept, be sure to focus on:

- Section 24.2, Energy Capture in Ecosystems
- Figure 24.5, The Dead Zone

Nutrient Cycles

Ecologists use the term *nutrient* differently than most people. To an ecologist, nutrients are the various atoms and molecules needed to build the body of an organism. Nutrients may or may not provide energy. Thus, the most important nutrients tend to be simple chemicals like CO_2, NO_2-, and K^+. Producers obtain their nutrients from the soil, water, or air. Consumers must eat either producers or other consumers. The transfer of a nutrient between organisms and the physical environment is termed a *nutrient cycle*. Nutrients may be difficult for producers to obtain, and shortages often limit the productivity of an ecosystem. The time required for a nutrient to cycle between a producer, the physical environment, and another producer depends on whether the nutrient cycles in an atmospheric cycle or a sedimentary cycle. In atmospheric cycling, the nutrient is a gas. It moves freely as the atmosphere circulates and tends to be available to almost all ecosystems. Nutrients that cycle in the atmosphere are rarely limiting. In the sedimentary cycle, soluble nutrients are transported by river to the sea, where they eventually become incorporated into marine sediments and become sedimentary rock. Over millions of years, these sediments may be uplifted, exposed to weathering, and eroded, allowing the nutrients to reenter the biosphere. Because this time interval is so long, these nutrients have the potential to become limiting factors. Phosphorus is the only major nutrient with a sedimentary cycle. Human activity commonly disrupts nutrient cycling, leading to surplus nutrients in some ecosystems and depletions in others. One of the earliest studies of nutrient cycling demonstrated that forest clear-cutting significantly accelerates the rate of nitrogen loss from forest soils. The practice of wastewater treatment often adds phosphorus to lakes and rivers, contributing to eutrophication and a deterioration of water quality.

For more information on this concept, be sure to focus on:

- Section 24.4, Nutrient Cycles
- Figure 24.9, The Sulfur Cycle
- Figure 24.10, The Phosphorus Cycle
- Figure 24.11, Altering Nutrient Cycles in a Forest Ecosystem

Tying It All Together

Several concepts presented in this chapter build upon concepts presented in previous chapters and may also be revisited and discussed in greater detail in the subsequent chapter, including:

Ecosystems

- Chapter 24 – Section 24.2, Exploitation

Energy Capture in Ecosystems

- Chapter 8 – Section 8.2, Photosynthesis: Capturing Energy from Sunlight

Nutrient Cycles

- Chapter 21 – in Section 21.4, Growth is limited by essential resources and other environmental factors
- Chapter 25 – Section 25.3, Changes in the Global Nitrogen Cycle
- Chapter 25 – Section 25.4, Changes in the Global Carbon Cycle

Practice Questions

Factual Knowledge

1. To continue to provide high-quality water for its residents, New York City decided to
   a. build desalinization plants along its Atlantic shoreline.
   b. construct water treatment facilities for its entire water supply.
   c. purchase the land bordering its watersheds to prevent development and pollution from further reducing water quality.
   d. construct rainwater capture basins to collect and store rainwater.
   e. all of the above

2. The greatest amount of energy is available at the ________ of a food chain.
   a. bottom
   b. top
   c. middle
   d. top and middle
   e. none of the above

3. The amount of nutrients available on Earth is
   a. increasing.
   b. fixed.
   c. decreasing.
   d. variable.
   e. none of the above

4. All life on Earth depends on ________ energy.
   a. solar
   b. heat
   c. chemical
   d. synthetic
   e. none of the above

5. Because of limitations on resources, organisms must ________ materials in order to survive.
   a. stockpile
   b. reinvent
   c. reintroduce
   d. recycle
   e. none of the above

6. Ecosystems are ________ because resources ________ move from one ecosystem to another.
   a. open; can
   b. closed; cannot
   c. one-way; can
   d. one-way; cannot
   e. none of the above

7. Ecosystem ecologists measure the
   a. diversity of species in an ecosystem.
   b. movement of energy and materials into and out of an ecosystem.
   c. natural resources available to a system.
   d. all of the above
   e. none of the above

8. Energy is recycled. (True or False)

9. Net primary productivity (NPP) of a system refers to the amount of energy
   a. obtained through photosynthesis.
   b. lost through cellular respiration.
   c. obtained through photosynthesis minus the amount lost through cellular respiration.
   d. stored as chemical energy.
   e. none of the above

10. NPP depends only on the amount of sunlight available. (True or False)

11. The majority of available nutrients in terms of biomass go to
    a. producers.
    b. consumers.
    c. decomposers.
    d. producers and consumers combined.
    e. none of the above

12. The primary difference between sedimentary and atmospheric cycles is that in ________ cycles the nutrient does not ________.
    a. sedimentary; leave the terrestrial environment
    b. sedimentary; leave the aquatic environment
    c. atmospheric; leave the aquatic environment
    d. sedimentary; enter the atmosphere
    e. none of the above

13. Acid rain results when humans put excess amounts of ________ into the atmosphere.
    a. phosphorus
    b. sulfur
    c. potassium
    d. hydrogen
    e. lithium

14. Match each term with the best description.
    __ trophic level
    __ ecosystem
    __ secondary productivity
    __ biomass
    __ nutrient cycle
    a. unit of measure of NPP
    b. another name for steps in a food chain
    c. all of the organisms and the environments in which they live
    d. rate of new biomass production by consumers
    e. cyclical movement of a nutrient between organisms and the physical environment

Conceptual Understanding

1. Secondary productivity depends on primary net productivity. (True or False)

2. Energy pyramids are used to represent energy transfer in an ecosystem because energy is ________ between each trophic level.
   a. gained
   b. lost
   c. conserved
   d. either conserved or gained
   e. neither lost nor gained

3. Rapid cycling of nutrients occurs in
   a. sedimentary nutrient cycles.
   b. atmospheric nutrient cycles.
   c. NPP.
   d. all of the above
   e. none of the above

4. Humans benefit from ecosystems because ecosystems provide
   a. buffers from natural disasters such as floods.
   b. maintenance of a clean water supply.
   c. climate moderation.
   d. all of the above
   e. none of the above

5. A pasture of 40 hectares has a biomass of 40 kg/hectare, or roughly 1,000,000 calories of energy. Cattle grazing on this pasture are sold for beef. About how many calories of energy from the pasture reach human consumers of the beef?
   a. 10,000
   b. 1,000
   c. 100,000
   d. 1,000,000
   e. 100

6. Phosphorus is a sedimentary nutrient. This means it
   a. cycles very slowly.
   b. never enters the atmosphere.
   c. settles as sediment in the ocean.
   d. all of the above
   e. none of the above

7. Decomposers such as fungi and bacteria are vital to the ecosystem because they
   a. are the primary organic recyclers.
   b. are the terminal energy consumers.
   c. keep populations in check.
   d. both a and b
   e. both a and c

8. Energy flow through an ecosystem differs from nutrient flow in that
   a. energy flow is linear.
   b. nutrient flow is linear.
   c. energy flow is cyclical.
   d. energy builds as it flows through the ecosystem.
   e. none of the above

9. The majority of the energy available in an ecosystem is actually lost as heat energy. (True or False)

10. Conifers are especially sensitive to acid rain because their needles literally rake moisture out of clouds. Because of acid rain, Mount Mitchell, the highest peak in the eastern Appalachians, has seen dramatic loss of pine forest. The damage on this mountain most likely stems from
    a. droughtlike conditions in the region.
    b. excessive sulfur in the atmosphere due to industrialization along the eastern seaboard.
    c. loss of biomass throughout the Appalachians.
    d. all of the above
    e. none of the above

11. Estuaries have one of the highest levels of net primary productivity because
    a. they are relatively shallow with good light penetration.
    b. nutrients transported by rivers enter the estuary before reaching the ocean.
    c. the action of the tides slows the deposition of nutrients into the bottom sediments.
    d. estuaries are often warmer than the adjacent river or nearby ocean.
    e. all of the above

12. Food chains of more than five trophic levels are quite common. (True or False)

Related Activities

- Investigate what is being done globally and by individual nations to address issues relating to acid rain. How is the United States attempting to curb acid rain? Write a one-page essay describing your findings.
- Identify at least two square-meter plots on your campus with different physical features like soil or exposure to sunlight. With the help of your instructor, devise a method to compare NPP for the sites. Write a brief summary of your findings, proposing a testable hypothesis that could account for any differences between the sites.
- Read or reread Rachel Carson's *Silent Spring*. Write a one-page essay that identifies which of her predictions have come true and talks about the progress we have or have not made in the areas she addressed.

Answers and Explanations

Factual Knowledge

1. **c** The high cost of alternatives helped New York City realize that protecting the natural system that provided clean water was cost effective.

2. **a** Producers capture solar energy. At each subsequent trophic level only about 10 percent of the energy is transferred.
3. **b** The fixed nature of nutrients means that they must be recycled.
4. **a** Producers convert solar energy to usable energy by way of the remaining trophic levels.
5. **d** Nutrient cycling is necessary for life on Earth.
6. **a** Energy and materials can pass from one ecosystem to another.
7. **b** By means of uniform measurements like biomass it is possible to make comparisons between ecosystems.
8. **False** Energy moves through an ecosystem.
9. **c** The "net" of anything is equal to that which was acquired minus the cost of operation to acquire it.
10. **False** Producers can't produce without adequate water and CO_2 and favorable environmental conditions.
11. **c** Decomposers take in more than 50 percent of available biomass.
12. **d** This characteristic makes sedimentary nutrient cycling necessarily slow.
13. **b** Excess sulfur is common in industrialized nations and causes acid rain.
14. **b** trophic level
    c ecosystem
    d secondary productivity
    a biomass
    e nutrient cycle

Conceptual Understanding

1. **True** Consumers can produce only if they have producers to eat.
2. **b** Only 10 percent of the energy is transferred between trophic levels.
3. **b** Atmospheric availability allows for more rapid nutrient cycling.
4. **d** With such benefits available, we should be responsible about our use of ecosystems.
5. **c** Only about 10 percent of the energy is transferred between trophic levels; thus, 100,000 calories are in the cows and available to human consumers.
6. **d** Sedimentary nutrients cycle slowly because they settle on the bottom of the ocean and do not enter the atmosphere. The upwelling of the ocean floor is needed for them to be released and made available for biological use.
7. **a** Decomposers are vital to the health of the ecosystem because they prevent nutrients from being tied up in one place for extended periods of time. Energy is not recycled; vast quantities have become fossilized without any appreciable impact on ecosystem function.
8. **a** Energy cannot be recycled.
9. **True** A large portion of available energy is lost as heat.
10. **b** Excess sulfur is a primary cause of acid rain.
11. **e** Estuaries are somewhat protected, often warmed, and enriched with nutrients from the nearby terrestrial ecosystems, conditions that support high NPP.
12. **False** Because only 10 percent of energy is transferred from one trophic level to the next, most food chains have only a few levels.

CHAPTER 25 Global Change

Getting Started

Below are a few questions to consider prior to reading Chapter 25. These questions will help guide your exploration and assist you in identifying some of the key concepts presented in this chapter.

1. Why has the harvest of large predatory fish like swordfish and shellfish like the white abalone been in decline for the past 40 years?
2. When land is transformed what does it become?
3. How does the current amount of wetlands in your state compare to the amount present in 1780?
4. What are synthetic chemicals, and why are organisms so poorly able to deal with them?
5. If nitrogen is a limiting resource, why have increased levels of nitrogen caused environmental problems?
6. What were the results of the Montreal Protocol?
7. In addition to carbon dioxide, what other greenhouse gases are serious environmental concerns?
8. Why is the conclusion that global temperatures are rising so controversial?

A Guide to the Reading

The following concepts typically give students the most difficulty when exploring the content in Chapter 25 for the first time. For each concept, one or more references have been identified that may help you gain a better understanding of these potentially problematic areas.

A Worldwide Change in the Environment

Human activity has substantially changed the biosphere in the past 1,000 years. Collectively, these changes are referred to as *land transformations*, when the land surface has changed, and *water transformations,* when aquatic biomes are affected. Abundant evidence documents these changes, and the total area affected is surprisingly large. Ecologists estimate that humans have substantially altered one-third to one-half of the Earth's land surface. Water transformation has been more difficult to document, but declines in the worldwide harvest of fish and shellfish are unprecedented in modern times. Ecosystems within these transformed areas have been dramatically affected. In the United States, wetland acreage, for example, has been in a continuous decline in every state for the past 200 years. When not directly lost, many ecosystems have been extensively modified to produce the goods and services needed by a growing human population. Humans are currently estimated to control 30 to 35 percent of the world's net primary productivity. Diverting such a significant portion of an ecosystem's NPP greatly reduces the amount of land and resources available to other species.

For more information on this concept, be sure to focus on:

- Section 25.1, Land and Water Transformation

Global Changes in Nutrients

The benefit of supplementing soil nutrients has long been known. In modern agriculture, the application of both synthetic chemicals, such as pesticides, and natural chemicals, such as phosphorus and nitrogen, has reached alarming proportions. Although it is almost 80 percent of the atmosphere, atmospheric nitrogen cannot be used directly by producers. The conversion to a usable form, called nitrogen fixation, is performed by certain species of bacteria. Currently, the amount of nitrogen fixation accomplished by human activities exceeds that of all natural sources combined. The consequences of changing the nitrogen cycle are far-reaching. When nitrogen is a limiting factor, NPP increases dramatically. Not all members of the ecosystem benefit, however. Those able to use the extra nitrogen may outcompete other species. One of the most common outcomes of nutrient enrichment is an eventual reduction in the number of species within the ecosystem. This is of concern because ecosystems are most stable when the number of species is greatest.

For more information on this concept, be sure to focus on:

- Section 25.2, Changes in the Chemistry of Earth
- Section 25.3, Changes in the Global Nitrogen Cycle
- Figure 25.4, Human Effects on the Global Nitrogen Cycle

Global Warming

Atmospheric scientists around the world are uniformly concerned about the increases in atmospheric carbon dioxide. Beginning in the 1700s, Earth's atmospheric levels of carbon dioxide have increased to levels that have not occurred naturally for the past 420,000 years. Carbon dioxide is transparent to the solar energy that reaches the earth, but reacts strongly with the heat energy released from the Earth's surface. Acting just like the glass windows in a car on a cold clear day, the heat energy is trapped, warming either the car's interior or the Earth itself. Average global temperatures have been slowly increasing for as long as reliable records are available. Temperature and rainfall provide the primary influences that dictate biome development. As global temperatures change, so will the patterns of atmospheric circulation and rainfall. Climatologists wonder if the new patterns of rainfall will continue to support agriculture in the areas where it now occurs. The soils that make agriculture possible have developed with thousands of years of climatic influence. There is a significant possibility that rainfall patterns might shift to areas with unsuitable soils. The potential for human impact alone is tremendous.

For more information on this concept, be sure to focus on:

- Section 25.4, Changes in the Global Carbon Cycle
- Figure 25.7, Atmospheric CO_2 Levels Are Rising Rapidly
- Section 25.5, Global Warming

Tying It All Together

Several concepts presented in this chapter build upon concepts presented in previous chapters and may also be revisited and discussed in the remaining Interlude, including:

A Worldwide Change in the Environment

- Chapter 21 - Section 21.6, Human Population Growth: Surpassing the Limits?
- Chapter 23 - Section 23.3, Recovery from Disturbances
- Chapter 24 - in Section 24.2, Human activities can increase or decrease NPP

Global Changes in Nutrients

- Chapter 24 - Section 24.4, Nutrient Cycles

Global Warming

- Chapter 20 - Section 20.3, Climate Has a Large Effect on the Biosphere
- In Interlude E, Individual actions can have a ripple effect

Practice Questions

Factual Knowledge

1. According to ecologists' estimations, the amount of land surface that has been altered by humans is
   a. about one-fourth.
   b. between one-third and one-half.
   c. less than one-eighth.
   d. about three-fourths.
   e. undetermined at this time.

2. Humans control approximately ________ percent of the world's net primary productivity.
   a. 10–15
   b. 20–25
   c. 30–35
   d. 40–45
   e. 50–55

3. Humans change how chemicals are cycled through ecosystems by
   a. adding synthetic chemicals to the atmosphere.
   b. releasing excessive amounts of natural chemicals through industrial processes.
   c. using chemicals to increase farm productivity.
   d. all of the above
   e. none of the above

4. The majority of available nitrogen on Earth is in the form of nitrogen gas, which cannot be used by living organisms. In order for it to be usable, nitrogen must be converted to
   a. nitrate.
   b. nitrite.
   c. ammonium.
   d. both a and b
   e. both a and c

5. The natural process by which nitrogen gas is converted to usable forms is called
   a. nitrogen fixation.
   b. synthetic nitrification.
   c. the Kjeldahl process.
   d. all of the above
   e. none of the above

6. Humans have added vast amounts of synthetically fixed nitrogen to the atmosphere. When this occurs, ________ increases, but ________ decreases.
   a. used nitrogen; available nitrogen
   b. species diversity; net primary productivity
   c. net primary productivity; species diversity
   d. available nitrogen; net primary productivity
   e. none of the above

7. Carbon dioxide concentrations in the atmosphere have risen dramatically over the past 200 years. The primary cause of this increase is
   a. high rates of natural decay.
   b. human burning of fossil fuels.
   c. agriculture.
   d. all of the above
   e. none of the above

8. The annual increase in carbon dioxide results from two activities: burning fossil fuels accounts for ________ percent of the increase, while deforestation causes the remaining ________ percent.
   a. 30; 70.
   b. 40; 60
   c. 50; 50
   d. 75; 25
   e. 90; 10

9. So-called greenhouse gases have been given this name because they
   a. trap the heat radiated from Earth's surface.
   b. are produced by plants.
   c. are necessary for plant growth.
   d. are similar to gases produced in a greenhouse.
   e. none of the above

10. Assertions that global warming is occurring are supported by observations that
    a. many northern plants have increased their growing seasons.
    b. several migratory bird species have shifted their range farther north.
    c. some butterfly species have been found farther north than ever before.
    d. all of the above
    e. none of the above

11. According to ecologists, humans must ________ in order to protect the Earth.
    a. note the damage that already has been done
    b. strive to live with the goal of sustainability
    c. reduce population growth
    d. realize the consequences of human activity
    e. all of the above

12. Many species are now ________ because of human impacts on the biosphere.
    a. extant
    b. extinct
    c. in hiding
    d. all of the above
    e. none of the above

13. Match each term with the best description.
    __ land transformation
    __ water transformation
    __ global change
    __ global warming
    __ sustainable living
    a. responsible use of resources by humans
    b. alteration of waterways by humans
    c. changes in land through human activity
    d. sum total impact of humans on Earth
    e. overall increase in Earth's temperature due to effects of greenhouse gases

Conceptual Understanding

1. In the Midwest, a high percentage of land is devoted to agriculture. There also has been an increase in large-scale hog lot operations. A sampling of streams throughout the region shows an increase in ________ due to fertilizer and animal waste.
   a. nitrogen
   b. carbon dioxide
   c. sulfur
   d. phosphorus
   e. none of the above

2. Vast tracts of rainforest have been clear-cut and burned. This practice has increased atmospheric carbon dioxide levels significantly because the
   a. forest provided a place for the carbon dioxide to be used.
   b. burning of the forest released a large amount of carbon dioxide into the atmosphere.
   c. grasslands that replaced the forests can't utilize as much carbon dioxide.
   d. all of the above
   e. none of the above

3. The major source of excess nitrogen entering the nitrogen cycle is from
   a. bacterial fixation.
   b. animal waste.
   c. synthetic fixation for use in agriculture.
   d. all of the above
   e. none of the above

4. Increases in productivity most likely cause loss of species because
   a. those species that increase in productivity outcompete other species.
   b. the causes of increased productivity in one species do not necessarily have the same effect in all species.
   c. there are inadequate resources available for productivity increases in all species.
   d. all of the above
   e. none of the above

5. Humans disturb chemical cycles by removing large amounts of chemicals from ecosystems. (True or False)

6. Which of the following is a potential result of global warming?
   a. increased ocean levels
   b. destruction of cities and island nations
   c. loss of species
   d. loss of habitats
   e. all of the above

7. Although little on Earth has been untouched by humans, our major environmental impact comes from
   a. land and water transformation.
   b. land transformation only.
   c. water transformation only.
   d. agriculture.
   e. none of the above

8. The draining of wetlands for building residential communities and for use in agriculture is an example of
   a. land and water transformation.
   b. land transformation only.
   c. water transformation only.
   d. land conservation.
   e. none of the above

9. Human impact does not affect climate. (True or False)

10. There is no way for humans to create a sustainable lifestyle on Earth. (True or False)

11. The greatest danger posed by chlorofluorocarbons (CFCs) is that they
    a. are toxic.
    b. decrease the ozone layer.
    c. alter the nitrogen cycle.
    d. cause global warming.
    e. none of the above

12. Nitrogen gas makes up 78 percent of the atmosphere, and thus nitrogen is readily available to plants and animals. (True or False)

Related Activities

- Write a one-page report describing how emissions standards are used to curb environmental pollution. Try to include statistics.
- Write a one-page proposal for how people of your generation can help move society toward sustainable living.
- Investigate how the United States compares to other countries in terms of environmental impact. Write a one-page report of your findings that could be presented to a local political group.
- Evaluate the evidence for and against global warming. Write a two-page essay explaining the position that seems most strongly supported by the evidence.

Answers and Explanations

Factual Knowledge

1. **b** Humans have modified between one-third and one-half of all land on Earth.

2. **c** Human control of 30–35 percent of net primary productivity leaves just 65–70 percent for all other species on Earth.

3. **d** Synthetic chemicals, industrial processes, and agriculture all alter natural chemical cycles.

4. **e** Living organisms can make use only of nitrates and ammonium.

5. **a** The natural process of nitrification is called *nitrogen fixation.*

6. **c** Productivity does not increase uniformly across species; therefore, some species are driven out.

7. **b** Increases in carbon dioxide in the atmosphere have been caused largely by the burning of fossil fuels.

8. **d** Current estimates suggest that about 75 percent of the carbon dioxide is caused by burning fossil fuels and 25 percent is caused by the reductions in photosynthesis that follow deforestation.

9. **a** A blanket of gases around Earth acts much like the glass of a greenhouse because it holds heat generated by Earth inside the atmosphere.

10. **d** Although many politicians and policymakers would like to overlook the evidence, it is mounting along the lines presented here.

11. **e** As the bare minimum, we are obligated to do all of these things.

12. **b** Human actions have significant consequences, such as causing the extinction of other organisms.

13. **c** land transformation
b water transformation
d global change
e global warming
a sustainable living

Conceptual Understanding

1. **a** Fertilizer and animal waste increase nitrogen in ecosystems.

2. **d** Balance is necessary. When one part is upset, the results spread to a much wider arena, affecting not only that ecosystem, but others as well.

3. **c** Agricultural practices have a vast impact on the nitrogen cycle.

4. **d** Resources are limited. An increase in a resource that benefits one species often causes another species (or several species) to suffer. An increase in resources may mean that one species outcompetes another, that resources are not equally available to all species, or that the increased resources do not have the same effect on all species.

5. **False** Humans disturb chemical cycles by adding synthetic and naturally occurring chemicals to ecosystems.

6. **e** Global warming may have diverse effects, including rising sea level and destruction of cities, habitats, and species.

7. **a** Humans have significantly altered land and waterways for their use and have profoundly influenced the environment.

8. **a** This practice has an impact on both land and water.

9. **False** There is very strong evidence indicating that humans are influencing long-term climate change.

10. **False** Though not easy, achievement of this goal is possible through increases in education, individual action, research, government policy, and pressure from the business world to develop ecologically sound practices.

11. **b** Chlorofluorocarbons (synthetic chemicals used as coolants) have caused a decrease in the ozone layer. This decrease is serious because the ozone layer shields the planet from harmful ultraviolet light from the sun.

12. **False** Although the atmosphere is composed of 78 percent nitrogen gas, plants and animals cannot use this form of nitrogen. Gaseous nitrogen must be converted to nitrate or ammonium.

INTERLUDE E Building a Sustainable Society

Getting Started

Below are a few questions to consider prior to reading Interlude E. These questions will help guide your exploration and assist you in identifying some of the key concepts presented in this interlude.

1. Under what circumstances is the use of a natural resource sustainable?
2. When is the production of oil expected to peak?
3. What is happening to the Ogallala aquifer?
4. What percentage of forested land remains in Kenya?
5. What percentage of the population of the United States has access to safe drinking water?
6. What is a greenroof?
7. How have satellites contributed to our understanding of environmental problems?
8. What is the Index of Sustainable Economic Welfare?
9. What pollutant does the Kyoto Protocol seek to reduce?

A Guide to the Reading

The following concepts typically give students the most difficulty when exploring the contents in Interlude E for the first time. For each concept, one or more references have been identified that may help you gain a better understanding of these potentially problematic areas.

The Current Human Impact Is Not Sustainable

Would you be surprised to learn that you may be living in what future historians might call one of the "golden ages of humankind"? It is possible that the lifestyles you and your classmates enjoy may never be duplicated. There may be too few natural resources remaining for future generations to live like you do. That's because many aspects of your lifestyle are not sustainable. This Interlude discusses sustainability, an action or process that can be continued indefinitely. Using natural resources at a rate that exceeds the rate of renewal and disrupting important features of an ecosystem, such as through pollution, are unsustainable actions and processes. A good example of unsustainability is how we use groundwater. A significant proportion of the water used for drinking, agriculture, and industry is pumped from underground sources. The removal of groundwater in many regions now greatly exceeds the natural rate of replenishment. As a result, some aquifers, the underground reservoirs where groundwater is pumped from, have dried completely, while others have dropped so low that pumping water has become very costly. If present trends continue, it's possible some communities and farmlands will have to be abandoned when their access to groundwater is no longer reliable. Deforestation represents another serious nonsustainable activity, particularly in the tropics, where forests are often converted to cropland and pasture. Some regions have lost 100 percent of the forests that were present just 1,000 years ago.

For more information on this concept, be sure to focus on:

- In Interlude E, Declining water resources are a serious problem
- Figure E.4, Declining Groundwater Levels
- In Interlude E, Global deforestation continues at an alarming rate
- Figure E.5, The Destruction of Old-Growth Forests

Sources of Hope for the Future

There are many reasons to be optimistic about the future of the environment. The number of people who are committed to solving environmental problems continues to grow. Much of the current growth is the outcome of education, which is helping many to reach a fuller awareness of the extent of these problems. Education also plays a role in how people form their attitudes about nature. Historically, natural systems were considered valuable only if they provided direct benefits to people. Alternative views emphasizing less direct benefits, like clean water or aesthetic values, are becoming more common. Although concerned about nature, many feel unable to make any meaningful contributions to the efforts to solve environmental problems. Yet individuals can make a difference. Few of us will have the impact of Wangari Maathai, winner of the 2004 Nobel Peace Prize for his environmental efforts with the Green Belt Movement in Kenya. But, as consumers, each time you make a purchase you have the opportunity to support companies that recycle, minimize the amount of throwaway packaging, or produce energy-conserving devices. Such seemingly small actions can have enormous effects when a large enough group of people become involved. The application of modern technologies is allowing ecologists, for the first time, to observe world-scale changes as they happen. Satellites, in particular, allow us to monitor the entire Earth in unprecedented detail. Unfortunately, most preliminary data is discouraging, documenting the damage that individual researchers had long described. However, because satellites can provide real-time data, those efforts that do work can be identified immediately and disseminated. Long-lasting improvements will never come about entirely through the efforts of concerned individuals and scientists. The participation of governments and business will also be critical. Governments have the power to mandate change through laws that regulate the actions of its citizens and protect its natural resources. For example, in the United States, provisions of the Clean Air Act have significantly reduced sulfur dioxide emissions and decreased acid rain. Governments also have the power to form treaties with other governments to address those issues that extend past national boundaries. International treaties have already been successful in reducing the emission of ozone-depleting CFCs. Finally, let's consider the contributions of business in forming a sustainable society. Currently, corporations consume large quantities of resources and generate large quantities of waste. There is, therefore, tremendous potential for corporations to reduce the environmental impact of their activities. Perhaps more importantly, as societies seek to establish sustainable forms of development, corporations can respond much more quickly than government with the new technologies that will be required. Some of the largest energy firms have made sizable investments in alternative and renewable energy sources. Fuel cells, solar conversion, and wind power are among these newer technologies. Moreover, investing in these technologies can often improve a corporation's profits.

For more information on this concept, be sure to focus on:

- In Interlude E, Sources of Hope for the Future

Tying It All Together

Several concepts presented in this chapter build upon concepts presented in previous chapters, including:

The Current Human Impact Is Not Sustainable

- In Chapter 20 – The Human Impact on the Biosphere
- In Chapter 21, The Tragedy of Easter Island

Sources of Hope for the Future

- Chapter 23 – Section 23.3, Recovery from Disturbance
- In Chapter 24, The Economic Value of Ecosystem Services

Practice Questions

Factual Knowledge

1. The Ogallala aquifer provides much of the water used in
   a. Texas.
   b. California.
   c. Kenya.
   d. the eastern United States.
   e. Canada.

2. Most experts predict that oil production will
   a. increase indefinitely.
   b. peak sometime between 2010 and 2020.
   c. continue at its present rate for several hundred years.
   d. be irrelevant as new energy resources will soon be developed.
   e. none of the above

3. The populations of ________ are the most likely to consume unsafe water.
   a. South America
   b. Europe
   c. Australia
   d. North America
   e. Africa

4. One of the more interesting aspects of the Green Belt Movement is that
   a. it has no staff or funds.
   b. it is responsible for planting more than 20 million trees in Kenya since 1997.
   c. more than 3,000 tree nurseries participate in the movement.
   d. all of the above
   e. none of the above

5. In life-cycle engineering, a company considers
   a. how to market products that have a multi-generational appeal.
   b. how to extend the useful life of a product.
   c. the total environmental impact of a product from manufacturing through disposal.
   d. how to reduce the cycle time of design-manufacturing-distribution.
   e. all of the above

6. The cap-and-trade system was initially developed to support the reduction of what pollutant?
   a. carbon dioxide
   b. nitrous oxide
   c. nitric oxide
   d. sulfur dioxide
   e. CFCs

7. *Subsidence* is the term used to describe the
   a. drastic losses in stock values for companies prosecuted for breaking environmental regulation.
   b. collapse of the land surface over a depleted aquifer.
   c. collapse of a fishery after years of overfishing.
   d. reduction in carbon trapping associated with the loss of forests.
   e. none of the above

8. Historically, the environment was valued for the resources it could immediately provide; people today often value nature
   a. simply because of its beauty.
   b. because they better understand the complexity of a natural system.
   c. because of the uniqueness of different ecosystems.
   d. because they have a better understanding of human dependence on natural systems.
   e. all of the above

9. One shortcoming of the Gross Domestic Product (GDP) is
   a. its inability to track gains and losses in the stock market.
   b. that it does not consider the social and environmental costs of the nation's economy.
   c. the value of imported goods are not added to the total.
   d. the United States maintains an annual trade deficit.
   e. all of the above

10. Which of the following power production technologies is nonrenewable?
    a. wind-powered turbines
    b. hydroelectric dams
    c. coal-fired steam turbines
    d. natural gas–fired steam turbines
    e. both c and d

11. The Kyoto Protocol requires
    a. reductions in sulfur dioxide emissions.
    b. the complete elimination of agricultural uses of DDT by 2020.
    c. reductions in CFC emissions.
    d. reductions in carbon dioxide emissions.
    e. an international ban on trade involving endangered species.

12. The world's largest producer of carbon dioxide has refused to participate in the Kyoto Protocol. (True or False)

13. The United States had harvested most of its old-growth forests by 1920; significant concentrations remained only in
    a. Florida and the Pacific Northwest.
    b. Ohio, Illinois, and Michigan.
    c. New Mexico.
    d. Louisiana and Georgia.
    e. Colorado and Utah.

14. Governmental intervention can often result in regulations that produce sustainable harvests and long-term stability for a resource; a good example is the
    a. lobster fishery in Maine.
    b. old-growth Douglas fir forests in Oregon and Washington.
    c. worldwide reduction in carbon dioxide emissions.
    d. all of the above
    e. none of the above

15. Match each term with the best description.
    __ subsidence
    __ aquifer
    __ groundwater
    __ greenhouse gas
    __ sustainable
    a. the sinking of surface land when groundwater level drops
    b. an action or process that can be continued indefinitely
    c. an underwater body of water
    d. water recovered from below the Earth's surface
    e. an atmospheric component that captures and retains the heat radiated from the Earth's surface

Conceptual Knowledge

1. How might the decline in oil production expected within the next 20 years impact life in the United States?
   a. Food production, which is heavily dependent on mechanization, may decline.
   b. Food transportation, which is heavily dependent on trucking, may decline.
   c. Pollution may increase as coal replaces cleaner-burning petroleum products currently in use.
   d. Electricity may be more difficult to obtain and will probably be more costly.
   e. all of the above

2. There is an extensive awareness in the United States that environmental problems exist; however, when asked to assess these problems most people
   a. vastly overstate their seriousness.
   b. somewhat overstate their seriousness.
   c. accurately state their seriousness.
   d. somewhat understate their seriousness.
   e. believe these problems to be limited to Africa.
3. The number responding favorably to the question "Should the environment be protected?" increased by over 60 percent during the 1980s. What best accounts for this change?
   a. increased television programming devoted to nature shows
   b. increased personal travel to foreign countries with environmental problems
   c. the concerted educational emphasis under way in schools, colleges, and elsewhere
   d. an increase in pet ownership and the increased awareness of nature that comes with it
   e. the general increase in societal compassion and responsibility
4. Single individuals can have an impact on today's environmental problems when they
   a. purchase energy-efficient cars and appliances.
   b. don't purchase disposable items.
   c. support businesses that produce and market "green" products.
   d. purchase electricity generated from renewable sources even though it may be more costly.
   e. all of the above
5. Individuals who install a greenroof experience many benefits. Which of those listed below would primarily benefit the environment?
   a. reduced heating costs in winter
   b. reduced dust
   c. reduced rainwater runoff
   d. reduced cooling costs in summer
   e. increased absorption of carbon dioxide
6. In what way is deforestation related to groundwater reserves?
   a. When trees are removed, the remaining vegetation uses less water and the reserves increase.
   b. When trees are removed, rain runs off rather than soaking into the ground and the reserves decrease.
   c. When trees are replanted, foresters cause excessive erosion that causes the reserves to become depleted.
   d. The pastures and farms that replace forested area have high water requirements.
   e. There is no connection between deforestation and groundwater depletion.
7. In addition to science, research in ________ can help build a sustainable society.
   a. archeology
   b. psychology
   c. economics
   d. anthropology
   e. history
8. After years of stable harvests, the Maine lobster fishery saw big increases. Should ecologists be concerned?
   a. Yes, because with increased catch, lobster prices will become more affordable.
   b. No, it simply reflects the inevitable outcome of sound management.
   c. Yes, such an increase reflects a change in the lobster carrying capacity that might have resulted from a disturbance in the ecosystem.
   d. No, cycles of this type are common and lobster densities will soon return to normal.
   e. Yes, increased harvests will attract more lobster fishermen and overfishing will be a potential problem.
9. The cap-and-trade system has one drawback and is criticized by some experts because
   a. some companies can make profits by reducing their emission of pollutants.
   b. the government does not have a very direct role in the process.
   c. the total cost of reductions was much smaller than anticipated.
   d. polluting factories can purchase additional allowances, and the air quality in a local region can actually decrease.
   e. all of the above
10. The key element in a greenroof is the
    a. double-paned glass panels.
    b. use of special green roofing shingles or metal that reflect ultraviolet light and reduce cooling costs.
    c. bank of solar panels that convert sunlight into electricity.
    d. use of roofing materials made from recycled items like milk jugs and concrete.
    e. matrix of soil and plants growing on the roof of a building.

Related Activities

- On a map of the southwestern United States, trace the path of the Colorado River from its origin to the sea. How far south does the river currently reach? Write a one-page summary of why the Colorado River no longer reaches the sea.

- Some energy experts are unconcerned about the apparent decline in oil reserves, believing that alternative forms of energy will soon be available. Using the World Wide Web or other sources, identify at least 3 energy alternatives and make a chart of the plus and minus features of each.
- Investigate this NASA Web site, www.nasa.gov/centers/goddard/news/topstory/2004/0624hanpp.html, which contains satellite images of NPP and worldwide estimates of the percentage of NPP used by humans. Write a one-page summary describing how this value varies among the world's nations.

Answers and Explanations

Factual Knowledge

1. **a** The Ogallala is the largest aquifer in the United States. It provides most of the groundwater used for agriculture in Texas, Oklahoma, Kansas, and Nebraska.
2. **b** Experts differ, but most interpret the continuous declines in new discoveries as an indication that almost all large reservoirs have been located.
3. **e** Almost every African country is currently challenged by water shortages and water-quality issues.
4. **d** Begun by Wangari Maathai with the planting of seven trees in 1997, the movement is now the major force in the efforts to reforest Kenya.
5. **c** In life-cycle engineering, companies monitor the total environmental impact of their product, often resulting in reduced impacts and increased cost savings.
6. **d** First proposed in the 1980s, the system reduced the cost of lowering SO_2 emissions from $12 billion to $870 million.
7. **b** In subsidence, the land surface over a depleted portion of an aquifer collapses inward, producing what are commonly called sinkholes.
8. **e** As a result of education, people today appreciate a wider variety of aspects of nature and are more likely to agree that protecting the environment is important.
9. **b** Tremendous costs, such as medical care related to air pollution, are associated with the nation's economic output but ignored when calculating GDP.
10. **e** Using coal or natural gas to produce steam for electrical generation represents a use of fossil fuels that cannot be renewed.
11. **d** Ratified by 141 countries in 1997, but not the United States, the Kyoto Protocol is an international treaty designed to reduce carbon dioxide emissions to below 1997 levels.
12. **True** The United States is the world's largest emitter of carbon dioxide, but has refused to participate in the Kyoto Protocol.
13. **a** By 1920 almost all old-growth forest had been removed and the land converted for other uses; only the forests of the Pacific Northwest and portions of Florida forest remained.
14. **a** Lobster harvests were stable from 1950 until 1990, and have increased since that time.
15. **a** subsidence
    c aquifer
    d groundwater
    e greenhouse gas
    b sustainable

Conceptual Knowledge

1. **e** Virtually every aspect of modern society is dependent upon the use of fossil fuels; the mechanization and transport of food are particularly troubling as few people now live in rural areas.
2. **d** While almost everyone is aware that environmental problems exist, many people believe these problems to be minor issues and are not motivated to change their lifestyles.
3. **c** Education plays a critical role in informing people about the scope of environmental problems.
4. **e** Individuals can express their environmental concern by being selective in how they spend their money. Businesses respond quickly to economic pressures and often change their practices.
5. **e** All the items listed are true benefits, but a–d benefit the builders of a greenroof directly; reducing carbon dioxide benefits the environment more than the individual.
6. **b** Surface runoff increases once mature vegetation is removed, increasing soil erosion, nutrient loss, and reducing groundwater replenishment.
7. **c** Increasing the knowledge in any field has the potential to benefit society, but the most immediate gains will come from economics, which can help us understand the true costs of the goods we produce and the benefits of modifying that effort.

8. **c** Sudden changes in population density usually indicate ecosystem disruption; in this case overfishing had reduced one of the lobster predators. Such change is usually detrimental in the long run.
9. **d** Since allowances can be sold, some experts have expressed concern that they may become concentrated in a relatively few locations that would then experience increased emissions.
10. **e** A greenroof is a "living" rooftop, providing space for plant growth that traps dust, absorbs carbon dioxide, and creates wildlife habitat.